THE NEURON AND THE MIND

This book, a companion to William R. Uttal's earlier work on macroneural theories of mind–brain relationships, reviews another set of theories—those based on microneuronal measurements. Microneural theories maintain the integrity of individual neurons either in isolation or as participants in the great neuronal networks that make up the physical brain. Despite an almost universal acceptance by cognitive neuroscientists that the intangible mind must, in some way, be encoded by network states, Uttal shows that the problem of how the transformation occurs is not yet supported by empirical research findings at the micro as well as at the macro levels of analysis. Theories of the neuronal network survive more as metaphors than as robust explanations. This book also places special emphasis on the technological developments that stimulate these metaphors. A major conclusion drawn in this book is that it is not at all certain that the mind–brain problem is solvable in the sense that many other grand scientific problems are.

William R. Uttal is Professor Emeritus (Engineering) at Arizona State University and Professor Emeritus (Psychology) at the University of Michigan. He was one of the pioneering researchers in computational modeling and is the author of numerous books and over 140 scholarly articles.

Books by William R. Uttal

- *Real Time Computers: Techniques and Applications in the Psychological Sciences*
- *Generative Computer Assisted Instruction* (with Miriam Rogers, Ramelle Hieronymus, and Timothy Pasich)
- *Sensory Coding: Selected Readings* (Editor)
- *The Psychobiology of Sensory Coding*
- *Cellular Neurophysiology and Integration: An Interpretive Introduction*
- *An Autocorrelation Theory of Form Detection*
- *The Psychobiology of Mind*
- *A Taxonomy of Visual Processes*
- *Visual Form Detection in 3-Dimensional Space*
- *Foundations of Psychobiology* (with Daniel N. Robinson)
- *The Detection of Nonplanar Surfaces in Visual Space*
- *The Perception of Dotted Forms*
- *On Seeing Forms*
- *The Swimmer: An Integrated Computational Model of a Perceptual-Motor System* (with Gary Bradshaw, Sriram Dayanand, Robb Lovell, Thomas Shepherd, Ramakrishna Kakarala, Kurt Skifsted, and Greg Tupper)
- *Toward a New Behaviorism: The Case against Perceptual Reductionism*
- *Computational Modeling of Vision: The Role of Combination* (with Ramakrishna Kakarala, Sriram Dayanand, Thomas Shepherd, Jaggi Kalki, Charles Lunskis Jr., and Ning Liu)
- *The War between Mentalism and Behaviorism: On the Accessibility of Mental Processes*
- *The New Phrenology: On the Localization of Cognitive Processes in the Brain*
- *A Behaviorist Looks at Form Recognition*
- *Psychomythics: Sources of Artifacts and Misrepresentations in Scientific Cognitive Neuroscience*
- *Dualism: The Original Sin of Cognitivism*
- *Neural Theories of Mind: Why the Mind–Brain Problem May Never Be Solved*
- *Human Factors in the Courtroom: Mythology versus Science*
- *The Immeasurable Mind: The Real Science of Psychology*
- *Time, Space, and Number in Physics and Psychology*
- *Distributed Neural Systems: Beyond the New Phrenology*
- *Neuroscience in the Courtroom: What Every Lawyer Should Know about the Mind and the Brain*
- *Mind and Brain: A Critical Appraisal of Cognitive Neuroscience*
- *Reliability in Cognitive Neuroscience: A Meta-Meta-Analysis*
- *Macroneural Theories in Cognitive Neuroscience*
- *The Neuron and the Mind: Microneuronal Theory and Practice in Cognitive Neuroscience*

THE NEURON AND THE MIND

Microneuronal Theory and Practice in Cognitive Neuroscience

William R. Uttal

Routledge
Taylor & Francis Group
NEW YORK AND LONDON

First published 2017
by Routledge
711 Third Avenue, New York, NY 10017

and by Routledge
2 Park Square, Milton Park, Abingdon, Oxon, OX14 4RN

Routledge is an imprint of the Taylor & Francis Group, an informa business

Library of Congress Cataloging-in-Publication Data
Names: Uttal, William R.
Title: The neuron and the mind : microneuronal theory and practice in cognitive neuroscience / by William R. Uttal.
Description: New York, NY : Routledge, 2016. | Includes bibliographical references and index.
Identifiers: LCCN 2015050181 | ISBN 9781138640191 (hb : alk. paper) | ISBN 9781138640207 (pb : alk. paper) | ISBN 9781315617480 (e : alk. paper)
Subjects: LCSH: Cognitive neuroscience. | Brain—Physiology. | Neurobiology.
Classification: LCC QP360.5 .U88 2016 | DDC 612.8/233—dc23
LC record available at http://lccn.loc.gov/2015050181

ISBN: 978-1-138-64019-1 (hbk)
ISBN: 978-1-138-64020-7 (pbk)
ISBN: 978-1-315-61748-0 (ebk)

Typeset in Bembo
by Apex CoVantage, LLC

. . . the relationship between brain and cognition is still only poorly understood. Great progress notwithstanding, neuroscience still cannot answer the "big questions" about mind and intelligence.

(Olaf Sporns, 2011, p. 179)

If psychological states are constructed, emergent phenomena, then they will not reveal their more primitive elements, any more than a loaf of bread reveals all the ingredients that constitute it.

(Lisa Feldman Barrett, 2011, p. 124)

To be brutally honest, scientists do not yet have even the remotest idea of how visual experiences—or indeed any other kind of experiences—arise from physical events in the brain.

(Stephen E. Palmer, 1999, p. 618)

The problem of consciousness is completely intractable. We will never understand consciousness in the deeply satisfying way we've come to expect from our sciences.

(Dietrich and Hardcastle, 2005, cited in Rakover, 2011, p. 1)

A review of the neuroimaging literature suggests that selective association between mental processes and brain structures is currently impossible to find.

(Poldrack, 2010, p. 754)

. . . the initial hope of associating each brain area with a particular function (Posner, Petersen, Fox, and Raichle, 1998) has not been realized.

(Fox and Friston, 2012, p. 408)

FOR MITCHAN

CONTENTS

PREFACE

The brain is arguably the most complex entity that has ever existed. It is made up of a vast number of cellular components (neurons) whose abundance compares with cosmological numbers. Recent estimates for the number of cells in the normal human brain (based on measurements by Ananthanarayanan, Esser, Simon, and Modha, 2009; Azevedo et al., 2009) are that there are about 86 to 100 billion neurons in the brain (along with possibly an equal number of glial cells). Astronomers, for their part, estimate that there are between 10 and 100 billion stars in a typical galaxy like ours. Thus, the numbers in a galaxy and a brain are of roughly the same order of magnitude.

Such a comparison is misleading, however. Although the components of brains and galaxies are both heavily interconnected, the natures of the respective interactive forces among their respective components are very different. Stars interact by means of a single, uniform, and well-described force—gravity—that declines uniformly with distance. Quite to the contrary, the brain's neurons are both locally and remotely interconnected with one another by what other estimates suggest may be thousands or even tens of thousands of synapses with idiosyncratic effects. Each of these synapses (or combinations of them) may produce distinctively different interactive effects on a neuron quite unlike the uniform influence of gravity between stars. Interconnections between the brain's neurons, therefore, are not simple or uniform nor are they likely to be as simple as the forces operating among a galaxy's stars. Thus, the combinatorial complexity of the brain's neuronal network is probably far greater than that of a galaxy!

Given this level of complexity as well as the microscopic nature of neurons and synapses, formidable obstacles exist to learning about the mechanisms by means of which tangible brain activity *becomes* or *produces* or *is* intangible thought. Nevertheless, in the past few decades there have been some extraordinary technical developments that allow us to begin to examine the brain in ways that only a few years ago would have been considered impossible. Many of these developments are just becoming known to both the general public and the scientific community.

Twenty-five years ago, I set out the goals of my future research program in an article entitled "On Some Two-Way Barriers between Models and Mechanisms" (Uttal, 1990). In retrospect, I would have titled this article somewhat differently. I now read much of what I was saying there as a critical appraisal of neuroreductionism—in particular, how various aspects of neurophysiological science produce or are related to cognition. This has been the central theme of my work and is extensively discussed in my recent and not so recent books (Uttal, 2001, 2005, 2009, 2011, 2013).

I argue here that none of these neuroreductionist theories proposed to explain cognition at the neuronal level has yet matured to the point that we can experimentally examine most of their postulates and implications. My primary goal in this book is to evaluate the current state of theory in cognitive neuroscience specifically at the level of the cellular (i.e., the neuronal) components of the nervous system. Unfortunately, there are some profound barriers to progress in this field. What has arisen in the place of solid data-driven biological-based theory, I believe, are largely metaphors or analogs driven by available instrumentation—an engineering field where astonishing progress has been made. An emerging generalization is that our theories are, too a much greater degree than we appreciate, creatures of whatever technology is available at any stage of history.

Therefore, it is a secondary goal of this book to survey the new technologies used by neuroscientists, to explain how they work, and to provide at least an initial conceptual foundation for the basic principles they highlight. I also want to consider what parts of the mind–brain conundrum they might help to solve and which may be completely intractable. Novel applications of these new techniques and the promise of even more exciting developments in the near future offer promise of major breakthroughs in our understanding of at least the anatomical and physiological aspects of brain, if not the "Holy Grail" of modern brain research: the answer to the question—how does brain activity account for mental activity?

Make no mistake, by the very act of applying some of these instruments, we come up against some of the most perplexing problems in modern science. Some of these problems are practical ones of dealing with the vast numbers of neurons that are involved. However, others are matters of deep principle that should be faced before any commitment can be responsibly made to such a grand enterprise as the proposed billion-dollar investment in the "Brain Activity Map" (Alivisatos et al., 2013), an approach whose goals are already coming into conflict with mathematical and physical laws that the brain must obey (Underwood, 2014).

In this context of questioning which goals of cognitive neuroscience are achievable using the best available technical developments and which are likely to elude our best efforts, there are several caveats that I should make explicit so that the comments presented here are not misinterpreted:

1. Nothing I say here about the difficulties arising in applying powerful new tools should be interpreted to mean that I am suggesting that we can determine at this point whether the mind–brain problem is either solvable or intractable. It is essentially a combined empirical and theoretical question that may not be answered fully for many years. Of one thing we can be sure—we are nowhere near even a tentative solution.
2. Should it turn out that the mind–brain problem is formally intractable, such a finding would necessarily have no special significance outside of the domain of the natural sciences. That we cannot understand something for practical reasons cannot be used to argue that there is anything other than materialist laws at work in the brain as well as in the cosmos. Nor does the absence of a complete explanation of a complex system such as the mind–brain mean that we are not able to manipulate that system in productive, practical ways.
3. Furthermore, no one can deny the possibility that some future discovery or some scientific genius will lead to an overarching theory of mind–brain relations and overcome what are apparently insurmountable current barriers to understanding. This book is only a contemporary evaluation of theory and technology underlying a particular kind of neuroreductionism, the kind we practice at the beginning of the 21st century. It makes no claims to the future beyond reasonable extrapolations from existing knowledge and identification of potential or real obstacles to progress.
4. The fact that there may be formidable technical and conceptual obstacles lying ahead should not be interpreted as a call for cessation of research at any level. We are learning much about single neurons and their properties in complex networks, and, at the very least, some network models or computer programs have led to useful simulations of human behavior of great value. The engineering of new tools that simulate human behavior will move on and whatever limits characterize current technology are likely to eventually, if not immediately, correct themselves.
5. Nevertheless, there are some fundamental barriers to understanding how the brain produces the mind that would be well to appreciate. To attempt to overcome some otherwise well-established physical law, such as the second law of thermodynamics, or the combinatorial explosion in computation in an effort to build a perpetual motion machine would be an enormous waste. We should not ignore the possibility that we might encounter similar physical or computational limits in cognitive neuroscience and beware of equally wasteful efforts to go beyond what we already know to be impossible.

In summary, there are two major questions that I seek to answer in this book. The first is—are overarching microneuronal theories of cognition possible? The second is—how does the available instrumentation constrain and dictate what kind of theories we develop? Should I make some slight progress in answering these two questions, I would consider my efforts worthwhile.

ACKNOWLEDGMENTS

As ever, I am extremely happy to acknowledge the support that I have received from my home unit at Arizona State University—the School of Computing, Informatics, and Decision Systems Engineering. I am grateful for the friendship of my colleague and friend Professor John Reich during the trying time this book was produced. I am also deeply grateful to my daughter, Taneil Uttal, who assisted me in the editing of this, my final book. However, most of all, it is my wife, Mitchan, to whom, in gratitude for her good sense, I dedicate this book once again.

1

INTRODUCTION[1]

1.1 Explanation

Of all of the scientific mysteries confronting our inquisitive species, none is more profound or challenging than understanding how the tangible brain can give rise to intangible thought. This has traditionally been known as the mind-body problem but, as we have become more sophisticated, referring to it as the mind–brain problem seems more appropriate. Despite the enormous amount of research being conducted that deals with the brain and the mind, it is almost universally agreed that we actually have made little progress toward resolving the overarching nature of this relationship. In place of creating anything approximating the grand theory of cosmological physics, we have learned an enormous amount about each domain separately but, unfortunately, piecemeal. Fortunately, knowledge about the anatomy and physiology of the brain is succumbing to powerful new research techniques, and psychologists have learned much about the ebb and flow of mental activity by observing behavioral responses.

What we have not done is to have made any substantial progress toward building a conceptual bridge between the psychological and neurophysiological domains. The reasons that we have not are multiple and complex; as a result investigators from many fields have debated whether the problem is in principle "solvable" or "tractable." On the one side are the optimists. To them, to "explain" how the brain makes the mind is just another scientific task, admittedly quite complicated, but still achievable within the rubric of conventional scientific inquiry. However difficult the problems may be, the obstacles to understanding them are just practical matters awaiting the development of a new method or technology. Most cognitive neuroscientists implicitly accept the ultimate achievability of the task and assume that however slow it may be in coming, eventually we will understand or explain how the brain makes the mind. To think otherwise would raise serious questions about the raison d'etre of cognitive neuroscience itself.

On the other side of the debate are the pessimists (or, if you prefer, the realists); their countervailing opinion argues that the problem is not just difficult in practice but is impossible in principle. That is, the complexity of the problem is such that there is no strategy or tool that we could adopt or invent that would ever lead to an analysis that would "explain" how the process of converting brain activity to mental activity takes place. Philosophers, combinatorial mathematicians, some behaviorally oriented psychologists, and even a few of those who would happily accept the title Cognitive Neuroscientist are now beginning to accept, implicitly if not explicitly, at least the possibility of impossibility.

The resolution of this dispute, of course, will ultimately come from empirical studies. Someday we may solve the problem and that will be that—debate over. Nothing proves the existence of a dimly seen phantom like Bigfoot better than robust evidence of its existence although absence of such evidence does not prove it does not exist. In this case, the "existence proof" for an overarching neuroreductionist theory would be the demonstration of our ability to construct high-level behavior or intelligence or sentience or cognition (or whatever it is that we may wish to call mind) from the properties of low-level neurophysiological components. Of course, even then the uncertainty of what amounts to an acceptable explanation would remain controversial.

This debate sets the stage for this book. In it, I strive to review and evaluate the current state of a particular kind of mind–brain theory—the kind that is based on the microscopic, cellular components (i.e., neurons) of the nervous system. These theories are designated as *microneuronal* as distinguished from those that are based on larger chunks of the brain—those designated as *macroneural*. I have already dealt with macroneural theories in a previous book (Uttal, 2016), to which this book is a companion, and will only deal with them casually in passing in this chapter.

To start, it would be useful to define exactly what we mean by a theory. For a number of reasons, however, categorizing existing theories is much more arbitrary than it may at first seem. Theories come in many kinds, each of which may emphasize different foundation axioms and postulates. A rough taxonomy of the major kinds of cognitive and cognitive neuroscience theories would include the following:

- Behavioral or Descriptive Theories: Theories of this kind are not meant to be reductive. That is, there is no effort to identify the cognitive or neurophysiological component parts that make up an observable behavior pattern. A pure behavioral theory, quite to the contrary, is only descriptive. It may be based on a mathematical formulation that allows us to predict the trajectory of a behavior or to represent the transformation between stimuli and responses, but there is no attempt to delve into lower-level cognitive or neurophysiological processes. Any *en passant* allusions to neurophysiological or inferred cognitive components are at best rough metaphors, in the absence of additional data. The mathematics itself, like behavior in a descriptive theory, is neutral with regard to specific underlying mechanisms—cognitive or neural. Thus, no matter how accurate are the predictions made by a mathematical model, unless there is additional data, it is not possible to produce a unique solution to the mind–brain problem from the formulation. There are always many alternative models that are sufficient to describe the behavior. Another way to say this is that behavioral and mathematical descriptions are *underdetermined*.

- Cognitive Reductive Theories: Theories of this kind are intended to be reductive but not to neurophysiological mechanisms. The underlying components are inferences implied by the behavioral data. Many theoreticians of this ilk hold the assumption that behavior is transparent (i.e., can be inferred from) the underlying cognitive process and that by careful experimental designs we should be able to parse out these inferred cognitive components. These cognitive components, modules, or faculties components have been designated as "hypothetical constructs" (MacCorquodale and Meehl, 1948), emphasizing the intrinsic difficulty of exactly defining them—a major handicap of such "top-down" approaches in psychological research. Again, these hypothetical constructs may allude in passing or by analogy to neural structures and processes, but it is impossible *in principle* for this approach to rigorously designate what these inferred components might be.
- Neuroreductive Theories: Because overt behavior and covert cognition are not able to lead us deductively to the underlying brain components that are necessary for mind, and because there is widespread agreement that all cognitive processes are in some ultimate sense brain processes, the search for the neural mechanisms of mental activity has motivated a vast amount of research. With the development of modern neurophysiological and neuroanatomical techniques, the search for the brain mechanisms of mind has greatly accelerated. There are two main threads of this kind of research, which has been alternatively known as physiological psychology, psychobiology, and most recently, cognitive neuroscience. These threads are respectively known as macro-neuroreductionism and micro-neuroreductionism.
  - Macro-neuroreductionism is based on data in which the cumulative, pooled, or summed actions of many individual neurons are emphasized. The details of what individual neurons are doing are lost since their neuroelectrical and neurochemical responses are additively lumped together in the eyes of the investigator. This kind of data provides a powerful impulse to develop theories that assume a kind of gross localization of the component neural processes on or in the brain. It also stimulates ideas about specialized cognitive roles for these macro-regions of the brain, that is, phrenological and neophrenological associations of places and cognitive phenomena.

- Micro-neuroreductionism is based on data obtained from experiments in which the activity and measurement of individual neurons is maintained either individually or in their participatory role as components of a neuronal network. There are three levels of micro-neuroreductionism within this part of this mini-taxonomy: (1) The action of single neurons has been associated with cognitive functions; (2) the action of relatively small, but computationally tractable, networks of a few neurons has been used as the basis of an intermediate level of theorizing; and (3) with the advent of supercomputers that can simulate the individual activity of billions of neurons has come a new form of micro-neuroreductionism approach in which the number of neurons involved in a theory can approximate the number of neurons in the brain. The ultimate form of a mind–brain theory, most cognitive neuroscientists probably agree, would ideally be found in the properties of such a network; however, it is still problematic whether we will be able to manipulate them in the way needed to "solve" the mind–brain problem.

A distinction should also be made between top-down and bottom-up theories in the context of this mini-taxonomy of theoretical types. The goal of top-down theories is to measure events at higher levels and then to determine by various methods the nature of the underlying neural and cognitive components that might embody the higher-level processes. The goal is, thus, to determine the underlying processes by drawing inferences from experiments on higher-level processes. This is the essence of top-down, analytical neuroreductionism. The underlying mechanisms are the unknown targets of this strategic approach.

Bottom-up theories have a different strategy; they attempt to synthesize higher-level systems from preexisting knowledge of the underlying components. This is essentially a synthetic approach to theory building. A classical example of the bottom-up approach to theory development was Watson and Crick's (1953) reconstruction of the genetic code for DNA from what was then known about the bonds between molecular structures. Their theory was embodied in a mechanical structure that depended on laws of chemical bonding in a way that opened up a new world in biological research.

Although both top-down and bottom-up methods seek to achieve the same goal—explain higher-level processes by lower-level ones—there is a fundamental difference in their basic approach. The analytical top-down method dotes on parsing a molar process into its components, whereas the synthetic bottom-up method aims to solve the problem by concatenating known lower-level processes. Both are susceptible to errors, of course. Top-down analytical methods are prone to mistake the sufficient for the necessary. Bottom-up synthetic methods are often more robust but suffer from incomplete knowledge of how the components may be interconnected.

All theories represent hopes for the future; however, each type of theory is constrained in what it can ultimately accomplish. Not only are there differences in the level of anatomical complexity at which a theory must operate but also there are major differences in what relationships should be measured in an experiment. Thus, the exact theoretical relation between a cognitive process and a neural response is not always clear even in the most highly correlated data.

Theories also differ in their methodological origins. A "theory" may be built on any one of the following relations:

- Statistical correlation between cognitive and neural responses
- The location of a neural mechanism of a cognitive process
- A mathematical description
- A physical model or simulation
- Reproduction by reconstruction, a rough verbal statement of some perceived metaphorical relationship
- The psychoneural equivalent of a cognitive process

The psychoneural equivalent is the actual neural mechanism whose activities are supposed to be indistinguishable from the cognitive process itself. In other words, the psychoneural equivalent is the neuronal embodiment of the cognitive process. Although defined in the different languages of cognition and neurophysiology, the different words may denote exactly the same thing. Determining the nature of the psychoneural equivalent is an extremely challenging task facing many impediments and obstacles; nevertheless, it is the holy grail of cognitive neuroscience. The most powerful interpretation—literal equivalence—in all of their properties—is in the manner expressed by "identity" theorists such as Place (1956) and Feigl (1958). In identity theory, cognitive and neural activities have the strongest possible relationship—the one is the other.

The range of theoretical issues with which I have dealt in my previous work on neuroreductionism can be further clarified by the following list of assumptions, properties, postulates, restrictions, and if you wish, biases that have guided my analysis:

1. Ontologically Physicalist: The prime postulate of modern cognitive neuroscience is the physical or material origin of mental processes. That is, whatever mind is, it is a manifestation or product of laws of the singular reality that accounts for all other events, things, and processes in our real world. In other words, cognitive neuroscience is fundamentally monist; it admits of no other

separable and distinguishable reality of the kind proposed by some philosophers. To accept any other (some kind of dualism) of a distinguishable reality difference between the mental and the physical would totally invalidate the science; it would require that we accept the possibility of forces that are not controlled by our experiments influencing those experiments. Such experiments, therefore, would invalidate the entire corpus of research in cognitive neuroscience.

2. Epistemologically Indeterminate: Despite the basic materialist ontological orientation of modern cognitive neuroscience, others believe that there are practical or epistemological constraints on what an empirical science can accomplish in the study of mind–brain relations. Some of these constraints have to do with the complexity of the system; some have to do with the inaccessibility of mental activity; and others arise from interpretive and logical errors. As a result of these practical problems, many proposed theories are little more than rough metaphors or prototheories rather than full-blown explanations. Thus, we may be in a situation in which despite being ontological monists, we are simultaneously doomed to be epistemological dualists and must study mind and brain separately.
3. Reductive: A reductive theory is based on the assumption that the properties of the lower level can, if appropriately manipulated, produce the properties of the higher level by concatenation. Nothing supernatural or additional happens between the two levels, just a rational flow of the processes and causes leading from one level to the other. The word "emergence" (by which is meant that new properties that have their origins in the lower level without causal connections or forces) is often used to denote the process but its use usually turns out to be a way of finessing the unknown without actually explaining anything. The radical reductionist, to the contrary, assumes that everything that is expressed at the higher level is at least implicit at the lower level. The task of cognitive neuroscience is to make explicit those properties that are implicit.
4. Bottom-Up Possibility: Bottom-up theories differ greatly from top-down ones. It is possible, in principle, knowing the properties and the rules of interaction (even if not practicable for reasons to be discussed later) to construct a bottom-up theory in which one reasons from the lower-level neurons to higher-level properties and thus to develop a "necessary," full, complete, and uniquely accurate theoretical explanation. An example of a bottom-up theory is how the human visual sensitivity to light of different wavelengths is accounted for by the differential absorption of photons of different wavelengths by the retinal receptor photochemicals. Another example of a bottom-up theory is the suggestion that Mach Bands (edge enhancements in visual perception) are accounted for by mutual lateral inhibitory interaction among retinal neurons. Another example of a successful scientific synthesis from a

totally different domain would explain how the microscopic structures of atoms account for the macroscopic behavior of chemicals.

5. Top-Down Impossibility: However, no matter how empirically robust, full, complete, and predictively accurate theoretical explanations they may be, top-down, neuroreductionist theories are not in principle robust. There are too many alternative, plausible, and possible "sufficient" explanations that can be generated from any given set of data. There is no way in which the underdetermined data of the macrocosm can be transformed into a unique "necessary" neuron-level explanatory theory. Top-down theories are useful for testing the plausibility of a prototheory but cannot distinguish between the unique "necessary" explanation and the array of "sufficient" ones that all fit the data equally well.
6. Neuronally Reductive: For the purposes of this book, the highest level of analysis is that of cognitively related behavior and the lowest level with which I deal is that of the specialized cells of the nervous system—the neurons. This instantiates a very particular postulate of the kind of neuroreductionist analysis dealt with here—namely, it is assumed, both explicitly and implicitly, throughout this discussion that any future neuroreductionist explanation of cognition is going to be framed in terms of individual neurons and their interactions at a microscopic level. It is further assumed that the identity and measures of the activity of individual neurons is necessarily preserved in any theory. This approach is distinguished from models that dote on macroneural "chunks" of the brain—a topic on which I have written extensively (e.g., Uttal, 2016)—and that have increasingly been rejected as a strategy that will lead us to the solution of the mind–brain problem. Other theoretical approaches, including the descriptive mathematical models commonly used in psychological theories, abound but are neither reductive nor neurophysiological no matter how well they correlate or predict the trajectories of the data.
7. A major principle, permeating everything I present in this book, is that available technology not only dictates what cognitive neuroscientists do in the laboratory but also guides our theoretical explanations. It is important, therefore, to distinguish between obvious and trivial metaphors and what are potential theoretical explanations. My strategy for doing this is to deal with the technology more thoroughly than is usual in a book of this kind.
8. Finally, to the extent possible, I remain an agnostic about the solvability of the mind–brain problem at the microneuronal level with which this book deals. As I discuss the empirical research on which my arguments are based, I doubt that I shall be converted from what is essentially a pessimistic position to an optimistic one, but at least some of the issues discussed will lead to a better understanding of what is involved in the search for the solution to the mind–brain problem.

The foregoing, then, are some of the assumptions influencing my analysis of microneuronal theories of cognition. To add substance to these initial perspectives, we need to review and evaluate the spectrum of proposed microneuronal theories and draw out their strengths and weaknesses. However, before inspecting individual theories and particular empirical results, there are some other general issues to be considered.

1.2 The Influence of Technology on Theory

It is an unambiguous cliché that science has always been driven by whatever measuring instruments are available to probe and measure natural phenomena. I argue here that the measuring instruments that can be marshalled by any generation of scientists play an essential role in determining not only what can be measured but also what theoretical inferences can and will be drawn. The history of science is replete with instances in which newly available instrumentation opened the door to new knowledge and deep understanding. It is also unfortunately true that, sometimes, whatever technology is fortuitously available can misdirect and even restrict scientific activities to a narrow path of action even when it is clear for other reasons that the path may be a dead end. It is all too easy to concentrate on what is available or easy to do rather than what is more fundamentally correct to do. Yet, as I have noted previously, *we do what we can do when we cannot do what we should do.* There is an obvious truism built into this aphorism, of course; one can only do what is doable and to ask one to do that which is not doable is a recipe for magical and dualistic thinking. The point is, nevertheless, that whatever tools are available at any point in science can exert a powerful influence not only on what interpretations are made of evidence but also on the nature of the evidence itself. That influence may mislead as well as illuminate as evidenced throughout the history of science.

A brief glimpse at the course of scientific history quickly makes clear that an array of influential instruments has guided the historical progress of the biological sciences and lately the cognitive neurosciences. The impact of Anton van Leeuwenhoek's (1632–1723) invention of the high-magnification microscope and its successors on biology in general and science in particular cannot be overstated. For neuroscience, the microscope and clever staining techniques (e.g., those developed by Golgi, 1906) clarified the cellular anatomy of the nervous system, opened the door to the study of individual neurons and, thus, provided the basis for theories emphasizing the role of neurons cum neurons in coding cognitive processes. Before the microscope, the brain was a homogeneous and undifferentiated "bowl of porridge"; afterwards it was a mesh of interwoven components.

Perhaps more than anything else, the discovery, measurement, and manipulation of electricity became the impetus for an enormous variety of scientific developments. In the present context, this revolutionary advancement in mind–brain thinking was based on two ideas. The first was that the nervous system's covert activity (as evidenced by the resulting overt behavioral responses) could be elicited by electrical stimuli—an idea famously known since the work of Luigi Galvani (1737–1798), in which a frog's leg was activated by an electrical shock generated by a primitive battery, the Voltaic pile. In addition to suggesting that there was an electrical basis of nervous action, the efficacy of electrical stimulation meant that very well-controlled stimuli could be used to generate well-defined responses in the nervous system.

The second, but far more difficult task, was recording the minute signs of nervous activity by measuring the electrical signals emitted by organic tissues and cells. This meant that the entire panoply of electrical measuring instruments developed over the centuries could be used to study the nervous system at several different levels of magnification. Although we now know that this electrical activity can be even more fundamentally explained in the language of ionic chemistry, specifically of the concentrations and flow of sodium, potassium, and chloride ions across the neuron's cell membrane, it was the doorway opened by electrical recording by the invention of some ingenious devices as the galvanometer and oscilloscope that made it possible to understand the operation of individual neurons. It is still not possible to unravel the complex activity of networks of many neurons, but progress looms on the horizon as new technological advances such as parallel-processing computers flood from engineering laboratories.

The point is that, to an immeasurable but very large degree, both experimental and theoretical progress in cognitive neuroscience, like all other sciences, has been driven by available instrumentation. That is, the instrumentation available at any moment in history largely determines our theoretical extrapolations and the inferences we draw from our observations. As a result, it is all too easy for us to be misled by tool availability to overvalue a particular line of theoretical thought to the exclusion of other more plausible, but empirically unavailable, alternatives. These biases also extend down into the empirical results themselves as instrument-based theory also feeds back to guide and direct our experimental designs and to select to which findings we will pay attention. Thus, for example, it is argued that many theories may be heavily (if not more so) influenced by the properties of whatever tool is available than some more germane but less obvious attributes of the phenomenon under study.

Consider the following example: The most popular of the current tools—the EEG and fMRI—used by cognitive neuroscientists mainly observe at the macroneural level. They record the activity of literally billions of neurons in the form of cumulative activity localized at particular places (on or in the brain) or occurring at particular times. This led to a number of prototheories of brain organization that were couched almost exclusively in the terminology of macroneural brain locations (using fMRI techniques) or temporal sequences (using EEG techniques) rather than the properties of microscopic networks. This has opportunistically led to theories of macroscopic mind–brain relations. In point of empirical fact, however, these cumulative, gross, macroneural measures may have little utility as reductive explanations of cognitive activity. It is becoming increasingly apparent that more valid explanations of mind–brain relations are likely to be framed in the concepts and terminology of the microneuronal elements and their interactions.

Of course, this microneuronal line of thinking approach itself does not exonerate any such theories from also misleading us; but, at least, there is also considerable logical justification for thinking that the microneuronal, neuron-by-neuron–based approaches are correct in principle and that the available macroneural approaches may be misleading us in practice. In short, given that we do not know the answers to most of the important mind–brain questions, I argue that we should be constantly aware that our instruments (and whatever kind of data they currently provide) can erroneously dictate our theoretical orientations.

Indeed, to a substantial degree, this is exactly what is happening (and has happened throughout the history of science). Current theoretical cognitive neuroscience thinking is heavily biased by the ease of availability of macroneural measuring devices. Brain-imaging studies based on fMRI technology impel us toward concepts of macroneural brain localization, and EEG studies drive us equally powerfully to theories of temporal "synchronization." Although these macroneural theoretical options are not always exclusive of microneuronal concepts, the amount of our intellectual energy committed to them is not justified by robust empirical evidence or reasonable logic—according to an increasing number of cognitive neuroscientists.

Thus, the caveat that the kind of instrumentation available exerts a powerful impulse to think in a certain way must never be underestimated. The ability to observe the response of an individual neuron has led to some fanciful "single cell" theories of equivalence between thoughts and individual neurons. On the other hand, according to most cognitive neuroscientists when pushed on the matter, the most likely level at which the nervous system engenders mind is that of the neuronal network, the microneuronal level at which there is virtually no empirical evidence to substantiate such an explanation. It is paradoxical that the least substantiated empirically is the most likely logically and widely accepted.

1.3 Barriers to Cognitive Neuroscience Theory Development[2]

Given the enormity of the challenge posed by the mind–brain problem, it is understandable that there are many obstacles and difficulties encountered when one attempts to develop an explanatory theory of the mind, whether it be cognitive, neural, or neuronal. The following discussion recapitulates some of these barriers and challenges that bedevil the task of the theoretician.

Levels of Analysis

A major barrier to theory building concerns the level at which a problem is being studied. By level of analysis, I refer to both the scale at which brain activity becomes mental activity and the scale of our analyses and measurements. In the present context, a microneuronal level of analysis entails investigating the responses of cellular and subcellular units such as neurons and synapses and their respective interactions. A macroneural level of analysis deals with signals that are inseparable pools or accumulations of these lower-level responses such as fMRI images or EEG recordings.

Theory development at a microscopic level is severely inhibited for several very practical reasons. Primary among these, we are overwhelmed by so much data that empirical-based theory is, for practical reasons, computationally unavailable to us. We have no way of both manipulating all of the necessary experimental details and measuring the vast amount of activity in the neuronal network whose collective but idiosyncratic states must in some sense correspond to cognitive processes. Macroneural approaches to theory are also deeply flawed. They trade the ability to measure for the convenience of pooled or mixed data, thus losing the very information—the detailed microstructure of the brain—that they would need to develop a microneuronal theory.

Thus, at both the microneuronal and macroneural levels, the possibility of developing an authentic, valid, overarching theory that truly speaks to the question of how the brain encodes mental activity is currently remote and is likely to be for the foreseeable future. For these reasons, I suggest that both levels of analysis are not promising foundations on which to build a theory of mind–brain relations. However, unlike the microneuronal level, macroneural experiments cannot only be conceived but actually implemented. This simple fact introduces a strong bias into the entire conversation. What follows is a more detailed discussion of some of the barriers that confront the cognitive neuroscience theoretician.

Complexity

It is a truism that the brain is an extraordinarily complex organ. Indeed, given the idiosyncratic nature of its interconnections, an argument can be made that it is the most complex entity in the universe. Regardless of which level of analysis one is considering, the combinatorial complexity of the brain's neuronal network stresses any possible explanation of its function to its limits for a relatively simple practical reason—there are too many neurons and too many interconnections involved in cognitive processes to be processed by any conceivable information processing machine. It surprises many how very few interacting neurons it takes to pose an intractable combinatoric problem. The complexity issue has been known for years. As long ago as the 1960s, Leon Harmon was demonstrating a simple three-neuron simulator that produced unpredictable results.

Complexity is a result of a number of factors. Obviously the sheer numerousness of the components involved in any realistic neuronal network that might be capable of instantiating a cognitive process is one factor. But, as we have already seen, even a modest number of a few interconnected components can produce combinatorial intractability. Furthermore, nonlinear neural mechanisms pose computational challenges that have not yet been solved by modern mathematics and remain among the most profound problems confronted by cognitive neuroscience.

Beyond sheer numerousness and nonlinearity, many early theories of neural nets did not scale up well. A simple neural network exhibiting, for example, learning with less than a hundred neurons might suddenly saturate with all simulated neurons responding simultaneously. This might occur when even a few more neurons were added or a slightly different network configuration introduced. The eventual collapse of the small neuronal network program in the 1980s (as discussed by Fodor and Pylyshyn, 1988), when attempts were made to scale up the number of emulated neurons, was probably an unavoidable outcome of this intrinsic instability. Although far more simulated neurons are involved in today's computational neuroscience models, these prototheories are also eventually subject to the curse of numerousness and combinatorial complexity.

Sheer numerousness, however, pales in significance when one also considers the idiosyncratic nature of the anatomy of the neurons and their interconnections that make up the brain. There are many different types of brain neurons and they are interconnected in a plethora of irregular ways by an abundance of synaptic variations. There are no shortcut methods to overcome this impediment to explanation that do not result in a massive loss of the very information that is presumably the psychoneural equivalent of cognitive processes. It has been argued that any proposed attempt to study the brain at this minute level is beyond both our technology and our mathematics (Lichtman and Denk, 2011).

Poor Definition of Cognitive Constructs

For science to achieve a high level of theoretical precision there is a profound need that the dimensions and properties of its variables be precisely defined. Yet, psychology is replete with poorly defined terms such as attention, thinking, emotion, and consciousness at a high level of abstraction and words such as learning, recognition, detection, retrieval, and conditioning defined more operationally. Many cognitive terms have multiple meanings that vary with the particular goals of the research. For example, Vimal (2009) tabulated 40 different meanings of the word "consciousness." As another example, I listed (Uttal, 2011) 41 different meanings of the word "learning." A further problem, however, is that none of these fluctuating cognitive terms need necessarily map directly onto neural mechanisms. That is, what is a fundamental elemental process or faculty to a psychologist may mean something very different to the brain. In short, our behavioral taxonomies and those of the brain may not be speaking the same language.

A novel way of emphasizing how this lack of isomorphism between cognitive and neural vocabularies affects our thinking has been provided by Oosterwijk et al. (2012). They concluded that the quest to associate specific cognitive constructs with particular brain regions has generally been unfulfilled. They, therefore, joined those who support the ideas embodied in an alternative hypothesis—namely that psychological constructs are encoded by a distributed pattern of responses executed by general purpose neural mechanisms. One implication of such a conclusion is that however much we may seek to modularize our psychological taxonomies (a tradition that goes back to Aristotle's faculty psychology) there is no reason to assume that the mental components are parsed in the same way by the nervous system. In Oosterwijk et al.'s words, "the brain does not respect faculty psychology categories" (p. 2110).

There are two ways to interpret this statement that are relevant to this discussion. First, psychological processes are not sufficiently well defined to be linked with the detailed responses of the underlying neural machinery. Second, as just noted, the brain is not necessarily modularized in the same manner as are cognitive processes. Either or both of these implications suggest that there is a fundamental mismatch between our psychological and neurophysiological languages and concepts. To search for localized psychological constructs in the anatomical brain, therefore, may be a search for a chimera.

Thus, if one seeks to find reliable equivalences or correlations, much less causal relations, between cognitive phenomena and neural responses, it demands an increased degree of precision in the definition of the key stimulus variables in an experiment. Unfortunately, different investigators often use different terminologies to specify what is actually the same cognitive process. One researcher's search for data backing up a theory of decision making, for example, may be operationally indistinguishable from another's effort to study attention. This is the inverse of the additional problem in which the same words may be used to define what may operationally be very different cognitive processes.

As a result, we have to take it as a given that cognitive neuroscience is as profoundly influenced by the vagueness of psychological language as it is by the availability of its tools. Clarifying psychological definitions and constructs should also be a high-priority task, especially for this science. Nevertheless, this important task is largely ignored by all but a few psychologists and cognitive neuroscientists. Currently, psychological terms are largely defined by the operations involved in carrying out an experiment. This can lead to serious confusions of definition both creating hypothetical constructs and ignoring real constructs.

Inaccessibility

Why should this imprecision of definition and measurement of cognitive processes exist? A major conceptual reason is that hypothetical cognitive constructs are not only inherently difficult to define but also impossible to directly measure. From any point of view, mental experiences are private, intrapersonal, and inaccessible psychological states that can only be experienced by the individual and are not directly sharable with others. Neither introspection nor experimental assays can provide us with direct evidence of the values, properties, and dimensions of mental phenomena. Our reasoning concerning "other minds" is, therefore, limited to analogies based on the Cartesian assumption that "Cogito ergo sum" and the analogous conviction that our fellows must be just like us. This means that we must use behavioral surrogates such as "percent correct" and "reaction time" as indirect measures of cognitive response, thus isolating us from the phenomena of primary interest to psychologists—our inner private cognitive responses.

Another reason that cognitive variables may be inaccessible is that they probably reflect the outcome of extremely complex multi-variate neural responses that cannot be characterized by single (or a few) measures typical of modern cognitive neuroscientific research. The net effect of inadequate definitions and inaccessible, underdetermined responses is to place conceptual barriers between the various levels of representation of cognitive activity and, thus, further limit the possibilities of developing a theory of mind–brain relationships.

The Search for Objectivity

Despite the difficulty in defining or measuring inaccessible cognitive entities such as thoughts, feelings, experiences, and perceptions, there has been a persistent historical effort to seek out objective "hard science" correlates—in particular, brain responses—of cognitive processes. The fundamental ontological premise of all such efforts has been the physicalist assumption that psychological activity is the outcome of brain activity although in some yet indefinable way. Few cognitive neuroscientists would disagree with this point. However, there is an epistemological corollary of this assumption that is a matter of great contention. Namely, can any measurement that is made of brain processes be theoretically transparent to cognitive activity? That is, are objective brain measurements capable of correlating sufficiently well with cognitive activity to serve as an "explanation" of that activity? Or, in other words, are the parameters of cognition inaccessible to science?

The main weakness of the a priori expectation that any objective brain response should be able to inform psychology about cognitive processes is that the multiple levels of brain coding mean that some objective measurements of brain activity, however empirically sound, are simply irrelevant to the mind–brain problem. Studies of the chemistry of individual neurons and synapses, microscopic investigations of structure, and single-cell recordings tell us little about the cognitively meaningful activity of what is increasingly likely to be widely distributed portions of the whole brain. It is not the molecular chemistry of neurons nor of synapses that might open the door to understanding how the brain works as a cognitive engine; it is more likely to be the aggregate information processing by a host of individual neurons whose collective (but not cumulative) states account for mind. This is a formidable barrier to building neuroreductionist theories of cognition.

Nevertheless, the search for objective measures of subjectively defined psychological states continues. In doing so it perpetuates the myth that observable, objective, neural brain responses representing cognitive processes are decodable. Even more seriously, the search for objectivity can bias our experiments such that the probability of false-positive results is inordinately enhanced.

False Analogies between Sensory, Motor, and Cognitive Processes

Another conceptual barrier encountered in the effort to formulate theories relating neural and cognitive processes is a false analogy being drawn between the indisputable empirical and theoretical successes in explaining sensory and motor codes, on the one hand, and the much more complicated matter of understanding the neurophysiological representations of cognitive processes, on the other. Peripheral sensory and motor signal transmission is characterized by a mainly unidirectional flow of information, relatively simple dimensional correlations between stimuli and neural responses, as well as a dimensional isomorphism between stimuli and perceived experiences. Perhaps most important is the fact that sensory and motor signals are anchored to the full gamut of physical measures and forces. Furthermore, the microanatomy of the peripheral sensory and motor pathways is much simpler than of those more intricate and interwoven central mechanisms that are presumably involved in cognitive processes. Sensory signals have a predominant afferent directionality, whereas motor signals have a predominant efferent directionality. Furthermore, sensory mechanisms often have a repetitive anatomy that is almost crystalline in nature, especially in invertebrates. As a result of the relatively simple anatomy and transmission roles of the sensory and motor pathways, it has been much easier to develop theories, indeed, nearly complete explanations, of transmission codes. The heyday of this kind of work was epitomized by the work of Hartline and Ratliff (1957) and Hubel and Wiesel (1965) and an army of sensory and motor neurophysiologists up to the present.

Thus, it is possible to ask and answer a question such as—what are the neurophysiological codes (i.e., the pattern of neural responses) used by the peripheral nervous system to transmit information about the magnitude or quality of an acoustic stimulus to more central regions of the brain? The physical dimensions of both the stimulus and the neural response are measurable in units for which the physical sciences have provided well-established values such as lumens and decibels on the one hand, and frequency and pattern of firing on the other. Furthermore, psychophysicists have developed powerful techniques for quantifying responses that require but the simplest of "Class A" discriminative responses (Brindley, 1960). These simplifying conditions do not hold for cognitive processes such as decision making or emotion, situations in which we are not really sure what the proximal stimulus really is.

Unfortunately, general disappointment was the main result when the same protocols and concepts were used and as false analogies were drawn between sensory and motor information transmission processes on the one hand, and cognitive activities on the other. These optimistic (and probably false) analogies persist to this day. In short, determining the neurophysiological basis of cognition is a much more challenging process than is the analogous task for sensory and motor processes. Progress in one does not necessarily portend progress in the other.

The Neurologizing of Psychological Language

A property of many psychological theories is the introduction of neurophysiological terminology into their discussions without adequate empirical linkage. Behavioral hypothetical constructs are supplemented by neural concepts and findings based on little more than functional analogies and borrowed language. Skinner (1950) referred to this use of unfounded neurophysiological concepts as a Conceptual Nervous Systems (CNS) approach, noting that:

> Many theorists point out that they are not talking about the nervous system as an actual structure undergoing physiological or bio-chemical changes but only as a system with a certain dynamic output.
>
> *(p. 194)*

What it appears that many investigators (e.g., Anderson, 2010; Johnson, 2011) are doing is to develop plausible, but highly speculative, "hypothetical constructs" that could, in principle, describe the behavior but which do not enjoy the support of empirical linkages between the psychological and neural mechanisms. The hypothetical mechanisms are then defined in neurophysiological terminology. At best, theories of this type are heuristics that are not likely to rise from prototheoretical hypotheses to robust neural theories. Perhaps we should be reevaluating the important insight about the Conceptual Nervous System offered by Skinner and distinguish between robust neurophysiological entities and "nicknames" for hypothetical constructs.

1.4 The Macroneural and the Microneuronal

It is vital to make a very important distinction at this point. The ultimate goal of the neuronal network theorist per se is either to develop a map of salient networks; or, if that is not possible, then it is to understand the general principles of networks such as those occurring in the brain. This is essentially a microscopic approach. The current goal of the cognitive neuroscientist, on the other hand, is to associate a particular mass of neurons (i.e., a location) with some cognitive activity by comparing the outcomes of experiments that involve both anatomical and functional methodologies. This is essentially a macroscopic approach. These two goals, it should be emphasized, are not the same and, for the moment, no bridge exists between the anatomical and the cognitive.

Each of these two approaches is challenged by technical and conceptual difficulties. Difficulties with the microneuronal approach (in which the integrity of the individual neuronal response is preserved) are counterbalanced by difficulties with the macroneural approach (in which the responses represent pooled or cumulated information that represents cognitive processes). Technologically, these two perspectives (the microneuronal and the macroneural) are very different in terms of their basic assumptions and the ease with which their respective difficulties can be overcome. Macroneural measures are cumbersome, noise ridden, and often quite arbitrary in their evaluation. They may require elaborate statistical analysis to extract localized regions of activity that might correlate with cognitive processes. This is the basic presumption forced on macroneural theorists by the instruments used to measure them. Although often expensive and time consuming, macroneural experiments are relatively easy to design to test their assumptions using such devices as the EEG and the fMRI. The main problem with these macroneural neural measures, however, is that they are cumulative responses pooling or summating the responses of individual neurons and their interconnections. Thus, much of the detailed information of microneuronal organization is lost. I have discussed the weaknesses of the macroneural approach much more completely elsewhere (Uttal, 2001, 2011, 2013, 2016).

Should the competing hypothesis—that the essential aspects of cognitive encoding are to be found at the microneuronal network level—be correct, then macroneural representation would be a deeply flawed approach to understanding mind–brain relationships. Rather than measuring macroscopic blood-oxygen levels with fMRIs as an indirect indicator of brain activity or summations of electrical activity from scalp electrodes with EEGs, we should be studying the microscopic action and interaction of neurons—and in particular the networks into which they are organized.

Empirical studies of microneuronal events are, on the other hand, much more difficult to implement experimentally. Most inhibiting to the microneuronal approach is that we have no specific knowledge of how many neurons are needed for a network to be "cognitively significant." It is likely that the number is very large—possibly as large as the number of neurons in the entire brain—but this is pure conjecture.

The Technological Basis of Macroneural Theories

In the spirit of the basic premise of this book that our technology strongly influences both our experimental results and our theories, I now briefly review the specific technology that has stimulated the current excitement about the macroneural approach using fMRI devices. The rush in recent years to develop macroneural theories and findings in cognitive neuroscience was presaged by the discovery of Nuclear Magnetic Resonance (NMR) by Rabi, Zacharias, Millman, and Kusch (1937, 1939). NMR was predicted by the emergence of quantum mechanics as a dominating theoretical approach during the 1930s. Although Rabi's group worked with gases enclosed in a Crookes tube, other investigators were shortly after able to reproduce the NMR phenomenon in solids and liquids.

The term NMR is somewhat self-defining. It is a phenomenon of the nuclear and subnuclear behavior of matter. The phenomenon depends on the magnetic properties of particles such as protons, which act as tiny magnets whose poles are in most situations randomly aligned. However, when a very strong magnetic field is applied to a substance (which may include organic tissues as well as gases and liquids), the magnetic orientation of all of the susceptible particles align in the same direction. If the strong magnetic field is released after a measured amount of time, the orderly proton magnetic alignments revert or relax back to their random state. If a small magnetic pulse stimulates the material during this "relaxation," the protons will emit radio waves at resonant frequencies that are dependent on the nature of the material of which the protons are a part and the nature of the small magnetic stimulating field.

These emitted radio signals can be picked up and after complex mathematical manipulations converted into three-dimensional spatial images of the material under study. The key aspect of this process is that different organic tissues will produce different patterns of radio frequency emissions. This was the major contribution of Damadian (1971), who first applied NMR techniques to imaging the human body. It was his inspiration that the NMR techniques developed in the laboratory for various gases and solids might be able to distinguish between normal and cancerous tissue in the clinic. Although his procedure was extremely primitive compared with modern equipment, clearly Damadian was the first to develop the imaging application of NMR. Subsequently, the brain-imaging strategy based on NMR physics revolutionized modern neuroscience. Nowadays, imaging methods based on NMR physics have largely replaced a wide variety of traditional methods involving surgery, trauma, stimulation, and electrophysiological recording.

The next step of consequence for cognitive neuroscience's role was the development of functional Magnetic Resonance Imaging (fMRI) by Ogawa, Lee, Kay, and Tank (1990). They showed how slight differences in the oxygen level of the blood could differentially determine the activity level of portions of the brain at different times and locations. The Blood Oxygen Level Dependent (BOLD) measures could thus be used to track functional brain activity as well as anatomical structure.

The earliest application of the fMRI technique to cognitive neuroscience is generally considered to have been published by Belliveau et al. (1991). (See the article by Kwong, 2013, for an interesting historical discussion of this groundbreaking experiment during which attempts were made to map out regions of the visual cortex that were selectively responding to visual stimuli.[3]) From this point in the history of this powerful tool, the influence of the fMRI technology on cognitive neuroscience cannot be overstated. Theoretically correct or not, fMRI technology changed the face of the science and created a zeitgeist in which the macroscopic measurements of brain activity were to dominate both experimental protocols and theoretical positions for years. Once again, the historical role of brain-imaging systems illustrates the strong impact that whatever technology is available to science can have on conceptual basis as well as its empirical findings. It was only decades later that some of the conceptual and technical deficiencies of the macroneural approach began to emerge.

Macroneural Theories—A Brief Review

Because of the current availability of particular kinds of instruments for measuring brain responses, the macroneural approach is by far the most popular for studying the functional (including the cognitive) role of the brain. The archetype of these instruments is epitomized by the fMRI system. It has many advantages, not the least of which is its noninvasive and benign nature. Currently, the number of articles published each year that deal specifically with comparisons of cognitive processes and fMRI brain images is difficult to determine because not all research using this method is related to what is currently referred to as the mind–brain problem (some have been directed quite successfully to related problems in medicine, neuroanatomy, or neurophysiology). Nevertheless, it is obvious that experiments comparing fMRI responses and cognitive processes have gone from virtually nothing in the early 1990s to tens of thousands in the 2000s (e.g., Bandettini, 2007; Aue, Lavelle, and Cacioppo, 2009).[4]

Because they are brain responses putatively correlated with cognitive processes, there is a compelling face validity presented to researchers by these new machines. The promise is that we are on the verge of a breakthrough in the profound and hitherto intractable mind–brain problem—how does the activity of tangible brain tissue produce intangible cognitive processes? The hope persists that we will eventually be able to meld what turns out to be an enormous database into a coherent, valid, and comprehensive theory of mind–brain relations. This face validity is, however, countervailed by a pervasive logical error that incorrectly assumes that because the mind is a brain process, *any* brain activity correlated with cognitive activity must be relevant to their interrelationship. At the least, it has been hoped that eventually correlations will be found between cognitive and neural responses that will serve as heuristics for plausible future theories of mind–brain relationships.

The conceptual postulate that dominates much of current macroneural cognitive neuroscience research is what we might refer to as the "traditional localization" concept. Essentially, this type of research is empirical but not directly theoretical; it is but a preliminary step in the acquisition of data that might help to build more comprehensive theories in the future. "Successful" experiments are those that are able to consistently associate particular localized peaks of regional activations with particular cognitive stimuli or task conditions. To the degree that these associations are robust and reliable, investigators try to add to a functional map of brain locations as their prototype theory. However, the basic idea that there are significant, localized activation peaks that selectively represent cognitive processes is now under attack. New experiments suggest that the "peaks" may be artifacts of sample size and statistical analysis procedures that prejudice the existence of the peaks. In fact, if these studies are correct, peaks of brain activity identified by brain-imaging equipment are artifacts in which our statistical techniques have created a false orderliness out of what are actually stochastic processes.

There, thus, prevails a highly questionable assumption guiding the research protocols in this macroneural mode of attack on the mind–brain problem, that is, that the inferred modular components of a cognitive process will map in some repeatable and neurophysiologically coherent way onto localized regions of the brain. As widely accepted as this hypothesis is, there have been persistent logical reasons and are now an increasing number of empirical reasons to question it. The psychological processes and phenomena for which localized representations are assumed are not necessarily dimensionally isomorphic with the brain's natural spatial layout. The cognitive processes are, it must also be remembered, themselves typically the instantiation of our experimental designs and not necessarily of any simple property of functional behavior. Whether they are divisible into the "hypothetical constructs" or intervening "modules or faculties" that correspond to specific anatomical regions or structures of the brain is uncertain.

The implications and inferences of hypotheses and theories must then be empirically tested to determine if they continue to hold more generally. However easy to put this essential step of the process into words, it is not that simple. Indeed, it is conversely true that when you are dealing with a system whose stimuli are at least multi-factorial and for which the triggering stimuli are obscure, whose responses are multi-dimensional and redundant, and for which there may be no direct relation of stimulus to cognitive response, the probability of finding any kind of a response that satisfies the a priori theoretical judgments of investigators becomes greatly enhanced.

The primary questions that may be answerable with brain-imaging techniques are variations on the theme of spatial localization; that is, *where* are the parts of the brain that are concomitantly activated with particular cognitive processes and how are they interconnected? This question can be approached from a number of different points of view, each with subtly different connotations. Although these differing connotations will become evident as I review the various kinds of research carried out that attack the problem of mind–brain relations at this macroneural level of analysis, it is useful at this point to tabulate some of the issues that have been considered by investigators using brain-imaging devices in their search for the foundations of a mind–brain theory at the macroneural level:

1. As noted, the archetype of the question asked by cognitive neuroscientists who use brain-imaging equipment and techniques is—*where* are the parts of the brain that become active when a particular cognitive process is under way?
2. Another level of inquiry is how, both in general and specific terms, are the salient brain areas interconnected at the macroneural level for cognitive operations? This was initially conceptualized as an *anatomical* question dealing with the nature of the interconnections (mediated by white, i.e., myelinated bands of axons) between regions. However, there is a functional analog that has generated much current interest—how are the parts of the brain *functionally* interconnected when particular cognitive processes are being carried out? Theories of this kind require much more elaborate analytical algorithms than simple tabulation of activated regions. However, they are really just another version of answers to the basic "where" question, one generalized to systems of interacting nodes rather than unique function-specific locations. The ultimate goal of this approach is to determine the connectivity among brain regions during cognition and then describe the properties of the resulting networks.
3. Classical physiological psychology has traditionally been aimed at associating anatomical brain mechanisms with particular cognitive processes by extirpative, stimulating, or recording techniques. However, in large part those kinds of research are beset by ill-defined and invasive surgical procedures and preexisting assumptions about regional functions. In large part, such techniques have been replaced by imaging techniques. Traditional work done using lesioning as the main tool (e.g., Kennard, 1955) has been much reduced, to be replaced mainly by fMRI techniques.
4. A classical problem that seemingly has been empirically resolved in the favor of distributed responses is the debate between those who argued that the brain correlates of cognitive processes were localized, function-specific regions and those who thought that the responses were distributed over broad multifunctional regions of the brain. The accumulating scientific evidence seems to increasingly support the latter conclusion. However, the details of this debate remain controversial, attracting the attention of a number of investigators.
5. Some cognitive scientists have proposed that current macroneural research with brain images will be able to resolve some purely psychological controversies. The question thus arising is—can neurophysiology inform psychology? This is a form of hypothesis testing that depends on testable neurophysiological postulates being included within what are otherwise purely psychological theories. The debate over the applicability of brain-imaging data rages on. (See the discussion between Coltheart, 2006, an opponent of the idea that any psychological theory controversy has *yet* been resolved by brain-imaging techniques, and Henson, 2006, who believes that they have.)
6. A major long-term and highly controversial issue in brain-imaging cognitive neuroscience is—can these techniques be used to read the mind, that is, to tell what a person was thinking about or perceiving by examining the fMRI

data? Although there has been some progress in selecting alternate sensory responses from among a limited set of brain images, much of this modest achievement seems to be attributable to topologically preserved peripheral sensory encoding. For example, visual stimuli are represented by retinotopic maps in the primary sensory area that preserve the topology of stimuli and, thus, may maintain accessible and useful information about the spatial pattern of a stimulus. This is not possible with the symbolic brain representations driven by higher-order cognitive processes that have no isomorphic relations between brain activity and those cognitive processes.

7. Many cognitive neuroscientists, heavily influenced by earlier psychological taxonomies, are trying to use brain-imaging devices to determine something about the localized brain mechanisms by which cognitive processes are carried out. For example, how do we learn? What brain changes occur in learning? A major unsolved problem is where and what is the engram? Others seek to understand the emergence of consciousness, attention, and other vaguely defined high-level cognitive processes by determining which brain regions are activated when these cognitive processes are manipulated. This is the main theme of much of brain-imaging research these days. However, there are compelling reasons to believe that this simplistic concept of seeking correspondences between cognitive processes and specific brain locations may be ill chosen.
8. Many technical issues occupy the time and energies of cognitive scientists, not all of which are aimed at the great question of how the mind emerges from brain processes. There is a continued effort to develop techniques to extract the best possible and largest amount of data from noisy brain images. Still other investigators are concerned with the technical matter of how we can pool or combine methods with low statistical power to produce higher-power experiments in order to yield more significant data—the meta-analysis approach. There remains an unresolved question when we use this method—can we plausibly combine experiments that are often varied in method and conceptualization? Depending on the actual signal-to-noise relations and whether a signal actually exists or is a manifestation of a stochastic system, this may also be an ill-chosen expenditure of resources.
9. Most generally, cognitive neuroscience is currently aimed at establishing the macroneural neural basis of cognition. In short, the main goal of this science is to provide some insights into the great question—how does the brain make the mind? To this overarching question, there is little in the way of either an answer or a satisfactory theory yet available. It is highly problematic whether we have even begun to answer this question given the likelihood that macroneural techniques such as the fMRI may actually obscure the critical microneuronal information that would answer this question.

Obviously, whereas some of these questions and goals represent issues that are of existential importance, others are of merely epistemic significance. Equally obviously, most of the cognitive neuroscience questions posed here are currently answerable. For those that can be answered, it is my hope that the subsequent discussions in this book will at least help to clarify the issues arising.

Like all other empirical approaches directed at a solution of the mind–brain problem, there are barriers and difficulties that impede progress toward a solution. Some are general to all cognitive neuroscience strategies while others are specific to the macroneural approach. One that falls into the latter class is the challenge faced by the pooled nature of the responses, that is, by the fact that any macroscopic measure is a composite of the individual responses of many individual neuronal responses. Currently, there is a prevailing opinion that it is the ensemble coding or states of the individual responses that in some ultimate sense must be understood if we are ever to make progress in unraveling the brain's role in cognition. Unfortunately for this approach, the pooled signals (the physical sum of many neuronal responses) actually have lost the critical information.

In such a situation, in which an exhaustive microneuronal analysis is beyond our technical capabilities, science tends to turn to available macroneural cumulative measures (e.g., the fMRI or the EEG) and effectively let nature do the accumulative analysis for approximate solutions to complex neuronal network problems. The expectation is that the critical microneuronal information will be at least partially preserved at the macroneural level. How the cumulative process works neurophysiologically is not completely understood, but it may involve summations of either local (e.g., Logothetis et al., 2001) or spike action potentials (e.g., Mukamel et al., 2005).

Currently, it seems likely that the salient microneuronal information is not, in fact, preserved in macroneural level measurements and, thus, the details of neuronal responses and interactions cannot be retrieved from cumulative measures such as fMRIs. The voltages and ionic currents that are the responses of individual neurons must follow the laws of physics as they are added and subtracted from each other. Once added or accumulated into pooled responses, basic thermodynamic principles argue that the original initial responses of a given neuron cannot be retrieved any more than an egg can be unscrambled. This constraint applies directly to the pooling of neuronal or hemodynamic responses directly or indirectly into signals like the fMRI or the EEG. Thus, there is an innate ambiguity about the origins of any macroneural signal based on accumulations of microneuronal activity—a huge variety of different configurations of microneuronal neuronal activities may produce exactly the same macroneural response. This means that scores based on differences between macroneural responses—for, example, the subtractive method—produce data that cannot discriminate between different microneuronal brain states. This issue is such an important matter that it has been continuously discussed for almost two decades (Van Orden and Papp, 1997) and is still in need of defense (Roskies, 2010).

The implication of the fact that the same macroneural responses can be produced by a variety of different microneuronal responses is that macroneural fMRI images are neutral with regard to their microneuronal origins. This is the fundamental weakness of the macroneural approach; the critical microneuronal information essential to understanding the transition from brain state to cognitive activity may have been irretrievably lost as a result of the pooling process.

The one possible exception to this neutrality of fMRI images is that information may be preserved concerning the respective locations of function-specific nodes, regions, activation sites, or locales on and in the brain—if they actually exist. Regardless of the truth or falsity of this fundamental postulate, finding the brain loci associated with cognitive processes has been the core of the larger portion of the research carried out with current fMRI equipment. Should it turn out that the macroneural representations of specific cognitive activities are actually valid and cognitive processes are localizable to particular nodes or activation sites on the brain, the location of the respective cognitive responses might be considered to be preserved information. However, even this most basic postulate of brain organization—localization—may not be empirically supportable.

Another closely related barrier to macroneural theory building is generically referred to as underdetermination. Underdetermination implies that there is insufficient information in the results available from an experiment or group of experiments to answer the salient questions for which the research was carried out. Some of this information was lost as a result of pooling as previously discussed—microneuronal information is not preserved in macroneural signals. However, it is more likely that most of it is due to the fundamental conceptual difficulty that the information necessary to answer a cognitive neuroscience question was never present in available macroneural measurements. Both behavior and formal mathematical models, for example, are underdetermined in that they do not contain enough information to uniquely determine the specific underlying mechanisms that produce the behavior (Moore, 1956). Indeed, neither behavior nor mathematics can distinguish among what may be an innumerable number of alternative possible and plausible underlying mechanisms.

Furthermore, additional experiments cannot always be carried out that will resolve this deficiency. Hilgetag, O'Neill, and Young (1996), for example, have argued on purely mathematical grounds that complex networks such as those proposed by Van Essen, Anderson, and Felleman (1992) for the primate visual system cannot be placed in a hierarchy of activation order because the outcomes are fundamentally underdetermined. Carrying out additional experiments leads, paradoxically according to Hilgetag and his co-workers, to an increase in the number of plausible mechanisms at a rate that exceeds the number of new experiments that can be designed to determine the response hierarchy. In short, simply collecting more data may not help to resolve issues of underlying mechanism if the data are underdetermined; indeed, it may exacerbate the problem.

Another way to describe underdetermination is closely linked to potential neural mechanisms. I refer here to the "many to one" constraint. If, for argument's sake, we accept the possibility of explaining behavioral observations in terms of macroneural mechanisms, then there is no reason to assume that a particular mechanism is solely capable of encoding any particular kind of behavior. It is far more likely that there are many different mechanisms that can equally well provide a satisfactory representation of the neural mechanisms associated with a particular behavior, that is, many neural mechanisms can produce the same behavior. Thus, a behavior is underdetermined with regard to the identification of a particular explanatory neural mechanism. It is also true that the inverse problem posed by the activation of a particular neural mechanism cannot be used as an indicator of a particular cognitive process. This is the problem of "reverse inference" highlighted by Poldrack (2006).

New and Relevant Data Arguing Against Macroneural Theories

An obvious difficulty that has to be acknowledged by cognitive neuroscientists is the undeniable (yet all too often denied) fact that the empirical foundation for robust macroneural theory building is still in an underdeveloped state. Comprehensive and accurate theories extrapolating from specific experiments to general laws are still relatively underdeveloped compared with the physical sciences. It is not yet certain whether this is a fundamental intractability or just the birth pangs of a new science like cognitive neuroscience.

If the ideal goal of a theory is to integrate data and abstract general principles, then there is an a priori requirement that the empirical foundation be robust, repeatable, and consistent; that whatever general principles are being expressed should actually be represented by dependable data. In the past several years, some investigators have reported experiments or analyses that challenge some of the foundation assumptions of macroneural theory building. In their most extreme versions, these assumptions include:

- That the brain is organized into regional subdivisions that encode or represent cognitive functions
- That these regions will exhibit a distinctive local neurophysiological response when a cognitive stimulus is activated
- That activation of one of these regions by, say, an electrical response will influence a relevant behavioral, cognitive, or neural response
- That reliable experiments can be reported in accord with robust experimental and statistical paradigms that permit replication of those experiments

I now review some of the most compelling arguments against these assumptions. Collectively these experiments suggest that serious questions are raised about macroneural theories at their most basic conceptual levels.

Thyreau et al. (2012) and Gonzalez Castillo et al. (2012)

Robust evidence supporting widely distributed brain states (as opposed to localized activations) of cognitive activity has now begun to appear in the literature (e.g., Gonzalez Castillo et al., 2012; Thyreau et al., 2012). Both of these groups of investigators presented compelling data that the apparent separation of brain responses into localized regions or functional nodes may be an artifact of inadequate sample sizes, arbitrary *p* values, statistical biases, and a prevailing presupposition of sparse localization. Gonzalez Castillo and his colleagues, for example, suggested that when adequate sample sizes (up to 500) were used to average brain images, localized activation areas in an individual subject tended to disappear, to be replaced by what is nearly a uniform overall activation of the entire brain. They did not exclude some relatively large regional differences but they concluded that:

> . . . under optimal noise conditions, fMRI activations extend well beyond areas of primary relationship to the task; and blood-oxygen level-dependent signal changes correlated with task timing appear in over 95% of the brain for a simple visual stimulation plus attention control task.
>
> *(p. 5487)*

Their results indicated that the more data that were included in the analysis, the greater was the extent of the distributed activity across the brain and less evidence existed for localized function. To support this result, Thyreau et al. (2012) reported a similar experimental result in which they also used a very large subject sample (n = 1,326) and found generally the same result—widespread distribution of responses across the entire averaged brain for even the most elementary cognitive process. They pointed out that when very large samples were used, even small responses began to achieve significance; this expansion increased until virtually the whole brain was responding significantly. The important implication of their work, like that of Gonzalez Castillo and his colleagues, was that the apparent segregation of the human brain into cognitively specialized regions associated with particular cognitive processes or brain activation nodes may possibly have been an artifact of inadequate sample size.

Oosterwijk et al. (2012) also supported the idea that the brain mechanisms of cognition must be more widely distributed in the brain than is currently held by localization theorists. From their point of view, the assertion that the neural mechanisms of cognitive processes are localized in the form of macroneural function-specific regions scattered across the brain was fundamentally incorrect. Specifically, they noted:

> Our results also emphasize the importance of examining distributed patterns of brain activation to understand mental states with different content

> rather than focusing on single regions . . . isolated regions may serve different psychological functions depending on what they are connected to during a given instance.
>
> *(p. 2125)*

The point of these studies is that the traditional search for localized macroneural regions mapping onto behaviorally defined psychological constructs is likely to be fruitless. The brain is not a system of isolated function-specific nodes; it is more likely to be a broadly, even universally, distributed system of neural mechanisms that are neither function-specific nor localizable to any particular region. Such a whole-brain metaphor is vastly different than the current one guiding much of macroneural thinking about how the brain is organized. In sum, the general import of this work is that the search for narrowly localized brain regions associable with cognitive processes may be sterile. The conflict between these emerging data and the search for localizable neural mechanisms may require a major change in whatever kind of theoretical model is being pursued.

Carp (2012a, 2012b)

How cognitive neuroscience data are reported is another general problem area that has recently come to the notice of the scientific community. The problems generated by the practical details of presentation and statistical analysis of fMRI measurements may be as potentially distorting to our theories as our choice of the level or kind of analysis at which we should work. We have recently been reminded of two serious problems concerning reporting of brain image-based experiments by Carp (2012a, 2012b). In the first of these two papers, he pointed out that the replicability of research in this field depends on adequate information being available concerning the design of an experiment so that the experiments can be reproduced. Yet, despite this universal necessity for replicability, when Carp analyzed 241 recent reports he found that "many did not report critical methodological details" (p. 289) in sufficient detail to permit replication. Not only was there a wide variety of procedures used for similar experiments, but also even the same data could be analyzed and reported in quite different ways, giving rise to diametrically opposed interpretations. Problems in reporting were identified by Carp concerning the design, the data acquisition, and preprocessing methods, as well as in modeling and display of the final results.

The problem of inadequate reporting, thus preventing adequate replication, is serious; however, another problem of perhaps even greater import to fMRI-based cognitive neuroscience research was also examined by Carp (2012b) in a follow-up article aimed specifically at the variability of the analytical methods that were used. He concluded that variations in method could lead to a cryptic increase in false-positive results. This hazard is exacerbated by the multiple modes in which data may be analyzed in an fMRI-based cognitive neuroscience experiment.

According to Carp, there are many alternative parameters of available analysis procedures with which to evaluate brain-imaging data. Depending on which subset of the parameters was chosen for a particular experiment, alternative conclusions could be drawn from a brain-imaging experiment. Carp suggested that ten is a plausible number of the key analytical parameters (including, for example, different kinds of normalization, filtering, and autocorrelation) as well as two or four alternative means of dealing with each of these analytical parameters. On this basis, Carp computed that there were 6,912 unique analytical "pipelines" that could be followed to a conclusion about the presence or absence of a significant effect. As Carp pointed out, however, not all of these pipelines lead to the same conclusion; indeed, there is a possibility of many competing Type I and Type II errors emerging from what were essentially the same measurements, depending on the investigators' more or less arbitrary choice of the particular pipeline that was used. In Carp's words:

> . . . results also varied considerably from one pipeline to another. Estimates of activation strength were highly variable across analytic pipelines; in regions of peak overall activation, significance estimates varied by 8 Z units.
>
> *(Carp, 2012b, p. 10)*

Although Carp does not assert that all experimental results vary across all pipelines, what he does argue is that "others varied widely from one pipeline to another" (p. 12).The fact that some pipelines, even the preponderance of them, were to some degree in agreement (a point made strongly in Carp, 2012a) does not mitigate the fact that there was an increased tendency for false positives to occur as a function of the variability in analytical methodology.

The conclusion to which we are drawn by this cogent critique and the concept of alternative pipelines is that variation in analytical method might be driving research conclusions (such as whether they are significant or not) to an unknown degree. Thus, great uncertainty is injected into even basic decisions about the significance or non-significance of the raw results. Should our scientific results vary with method and not converge for different methods, they would not inspire much confidence in their validity as measures of the underlying neural reality.

Similar points have been made by Simmons, Nelson, and Simonsohn (2011) and Bakker and Wicherts (2011) for purely psychological studies. Both groups also noted that there is such enormous flexibility in the way we collect and analyze data that spurious false positives are encouraged. They emphasized several additional factors as contributing to poor decisions that lead to error: (1) ambiguity about the basis for the $p < .05$ criterion, (2) the arbitrariness of when to terminate data collection, and (3) simple calculation errors. Bakker and Wicherts, for example, stated that as many as 33% of the 281 reports they reviewed either misreported their findings or made an error in calculation.

Horvath, Forte, and Carter (2015a, 2015b)

The general idea behind macroneuronal theories of cognition is that by appropriately activating some aspect of cognition, we can localize regions of the brain that are in some way responsible for representing that cognitive process. There is, however, a related idea that is based on the inverse of this idea—namely that we can stimulate certain regions of the brain to manipulate, influence, or evoke specific cognitive processes. In the former case, we are searching for the brain locations associated with predefined cognitive states. In the latter case, we are assuming that certain brain areas are related to certain cognitive states and then attempting to manipulate those cognitions by stimulating the brain locales with electrical stimuli. Both approaches are based on the same fundamental premise—namely that cognitive processes are mediated by specific macroscopic brain locations.

Work using transcranial Direct Current Stimulation (tDCS) has blossomed in the past decade with suggestions that electrical stimulation of localized regions of the brain can have profound effects on cognitive functions and even psychiatric problems. Dozens if not hundreds of reports now purport to document the psychological effects of electrical stimuli directly applied to the skull. Like many other enthusiastic applications of new technologies, it now appears that what seem to be a plethora of positive effects may actually be little more than random noise. A team led by Jared Horvath has reviewed and statistically analyzed large samples of tDCS experiments that implemented this localization-by-stimulation protocol. In two articles (Horvath, Forte, and Carter, 2015a, 2015b), on the basis of an extensive statistical analysis of the germane literature, Horvath and his colleagues concluded that there are *no* reliable effects of this kind of electrical stimuli on neurophysiological and behavioral measures chosen from a diverse sample of different types of experiments. Positive effects seemed to be balanced by negative reports over a wide range of cognitive experiments. The single exception to this general result was that motor twitches varied systematically in amplitude with tDCS stimulus amplitude. This is hardly a surprising result given that the stimuli were being applied over motor areas of the brain.

Horvath, Carter, and Forte (2014) went on to suggest that this poor reliability could be accounted for by five types of inadequate controls:

1. Intersubject variability
2. Intrasubject variability
3. Absence of sham stimulation and blind analysis techniques
4. Motor and cognitive interference
5. Controls for physical nature of the direct current stimuli

When examined in the light of the equally poor controls and variable results obtained with the fMRI approach, this critique speaks strongly to the idea that macroneural theories based on localized brain regions associated with cognitive processes, so widespread in current cognitive neuroscience, are inadequate.

Barrett (2015) and Barrett and Satpute (2013)

Another group that has been recently publishing research that is in contradiction to the idea that there are specific locations in the brain that encode or represent specific cognitive processes is led by Lisa Feldman Barrett of Northeastern University. She summarized this group's conclusions about the localization of emotions in an opinion piece in the *New York Times* (Barrett, 2015) in the following way:

> . . . [we] collectively analyzed brain-imaging studies published from 1990 to 2011 that examined fear, sadness, anger, disgust, and happiness. We divided the brain into tiny cubes, like 3-D pixels and computed the probability that studies of each emotion found an increase in activation in each cube. . . . Overall we found no brain region was dedicated to any single emotion. We also found that every alleged "emotion" region of the brain increased its activity during nonemotional thoughts and perceptions as well.
>
> *(p. 10)*

As well as:

> Although it has long been assumed that emotional, social, and cognitive phenomena are realized in the operations of separate brain regions or brain networks, we demonstrate that it is possible to understand the body of neuroimaging evidence using a framework that relies on domain general, distributed structure function mapping.
>
> *(Barrett and Satpute, 2013, p. 361)*

Button et al. (2013) and Uttal (2013)

Recently, a general appreciation that not all was well with imaging as the basic tool of cognitive neuroscience research—fMRI—has emerged. Two studies (Button et al., 2013; Uttal, 2013) pointed to statistical problems with the reliability of brain-imaging studies created by small sample size. The problem was that because of the expense of running subjects, many experiments were deficient in statistical power—a correlate of inadequate sample size. This permitted many statistically significant studies to be interpreted as having demonstrated positive results when, in fact, they had not.

Button et al. pointed out that low power led to "overestimates of effect size and low reproducibility of results" (p. 365) in general and of brain-imaging studies in particular. This meant that "the chance of discovering effects that are genuinely true is low" (p. 366). Beyond statistical formalities, this means that data from experiments will be highly variable even though individual experiments might be significant. Uttal showed that this lack of reliability was permeating the entire cognitive science application of fMRI machines by comparing a variety of experiments. The general effect was to produce variability and inconsistency despite what must have been acceptable levels of significance.

By 2005, the problem had become widely recognized and resulted in such hyperbolic statements as "Neuroscience research gets an "F" for reliability" in a *Scientific American* blog, "Brain science is drowning in uncertainty" in *New Scientist*, and "Brain-imaging is often wrong" in *Vox*, a "general news site." More seriously, submissions of this kind of research to various journals were down, especially if they dealt with comparisons of cognitive and imaging data. Efforts to overcome the technical and statistical problems resulted in grants to fund a "Center for Reproducible Neuroscience" at Stanford University. Although there may be some corrections to be made in the way data are collected and analyzed, it is not certain that further investment with what otherwise may be a flawed approach to the study of brain imaging carried out at the wrong level of analysis as a pathway to understanding the mind–brain problem is justified.

1.5 The Rise of Distribution—the Fall of Phrenological Localization

That we would be able to identify localized brain regions encoding specific cognitive functions was the hope of the early work using fMRI systems. This led to what some of us (e.g., Uttal, 2001) described as a neophrenological orientation to our theories in which a narrowly localized area of the brain was associated with poorly defined cognitive processes. However, increasingly, leading investigators such as Barrett (2015), Fox and Friston (2012), Button et al. (2013), and Poldrack (2010) have pointed out that the goal of assigning localized brain regions to cognitive processes has not been and possibly cannot be achieved. Fox and Friston look back over two decades in which a flood of papers sought to relate macroscopic brain locations and cognitive processes and concluded:

> Over the past twenty years, neuroimaging has been the predominant technique in behavioral and cognitive neuroscience. The volume of papers and number of fields it pervades are unrivaled. Despite this, it is curiously difficult to summarize its achievements in general terms. The simplest attempts falls [sic] back on two principles that shaped brain mapping at its inception; namely functional *segregation* and *integration*: Neuroimaging has established functional segregation (the segregated or modular deployment of functional specialization within brain regions) as a fundament of brain organization. . . . However the initial hope of associating each brain area with a particular function (Posner et al., 1998) has not been realized. While it is true that notions like the "motion sensitive center" and

> "fusiform face area" are part of common imaging parlance, the functionally informed labeling of all but the smallest portion of cortex remains elusive. Indeed people now prefer to talk about processing hierarchies, intrinsic networks and default modes that have no clear association with discrete cognitive processing components.
>
> *(p. 408, italics in original)*

Fox and Friston argue that the field has changed or is in the process of changing and we should no longer attempt to make simple associations between brain locales and cognitive processes. Although we are likely to continue to use brain-imaging equipment, they suggest that future work will increasingly be concerned with the anatomical organization of brain exemplified by the search for the connectome (e.g., Sporns, Tononi, and Kotter, 2005), the effort to determine how the brain responds when it is cognitively inactive (e.g., Fox et al., 2005), or, more generally, "a more physiologically and anatomically informed approach" (p. 408). Fox and Friston are not alone in drawing this conclusion. Poldrack (2010), another prominent investigator, concluded that "A review of the neuroimaging literature suggests that selective association between mental processes and brain structures is currently impossible to find" (p. 754).

In other words, the field of cognitive neuroscience is gradually changing from one that sought neural codes for cognitive processes to one more concerned with the properties of neural networks per se. It seems to be that neuroimaging is diminishing its activities related to cognition in favor of research that is more concerned with anatomy and physiology of networks. The reasons for this are clear:

- The solution to the mind–brain problem is more likely to be solved at the level of microneuronal networks than of pooled macroneural responses.
- Therefore, the technological methods and tools we currently have available to study the problem (most notably the brain-imaging technology) are inadequate and inappropriate for solving the mind–brain problem.
- Critical information is lost when neuronal responses are pooled or averaged.
- Cognitive processes are notoriously underdefined.
- Cognitive processes are underdetermined with regard to underlying neural mechanisms.
- Statistical methods cannot handle the "low power" experimental designs resulting from inadequate subject sample size.
- The problem of relating cognitive processes and neuronal components or networks is far more difficult than is generally appreciated.

What a terrible waste of resources resulted from the fact that early critics of brain imaging as a means of studying the neural basis of cognition were not attended to (Uttal, 2001; Van Orden and Papp, 1997).

1.6 Microneuronal Theories—A Brief Introduction

The main handicap of the macroneural approach to theory building is the irretrievable loss of critical information about the states and responses of neurons and synapses because of response summation. Likewise, the microneuronal approach is similarly obstructed but by what many consider to be the antithetical problem. Rather than having lost the necessary data, the difficulty facing microneuronal approaches to theory building is a glut of data. And, what a glut it is! The sheer number of neurons that must be involved in the representation of even simple cognitive processes is almost certainly in the billions. Furthermore, the number of relevant neurons pales in light of the number of salient synapses between neurons for which we must account. Even more detrimental to our hopes of building a microneuronal theory is that the nature of the interactions between neurons is not constrained to a single forcing function. In its stead, each neuron interacts with its neighbors by means of an idiosyncratic set of connections encoding a myriad of forcing functions. No universal rule of interaction comparable to the force of gravity or any of the other attractive and repulsive forces of physical nature operates in the same way between all of the components of the neuronal network to simplify our computations. This means that progress in developing microneuronal theories must be based on a brute-force analysis in which the individual responses of each and every neuron must be considered.

As we see later, this situation is only partially ameliorated by the advent of supercomputers, in which the number of logical units is becoming comparable to the number of brain neurons. If the microneuronal approach is correct, no matter how large the computer, it will still be necessary to take into account information about the states of the individual neurons. This means dealing with the neurons individually and combinatorial issues immediately become salient. It would still be necessary (if we succumbed to the poor judgment of actually trying to execute the kind of Gedanken experiment proposed by Martin, Grimwood, and Morris, 2000) to not only measure the state of each neuron but also to load the system with individual initial conditions for each neuron. Clearly, it is not just a matter of measuring N (where N, the number of brain neurons, is a large number itself) but also of evaluating the combinations of N that may result in data explosions such as N! (where N! is N factorial, a much larger number for even small Ns).

I hope it is not too pessimistic to suggest at this early point in this book that there is little hope even in the distant future for a full-scale theory at the microneuronal level. One must also keep in mind that Moore's (1965) doubling law of computer packing density (a surrogate measure of raw computer power) grows far slower than any of the functions that designate the possible combinatorial growth of complexity. Therefore, increases in computer power cannot be expected to overcome the innate intractability of the microneuronal approach to the mind–brain problem.

1.7 Interim Summary

The generic question asked in this chapter is—is it possible to produce an overarching theory of the process by means of which mental processes are encoded or represented by neural mechanisms? That they are, albeit in some so-far inexplicable manner, is the foundation postulate of modern cognitive neuroscience. Without such a foundation postulate, none of the extensive empirical research now being carried on would have any psychological meaning. Nevertheless, although we have had millennia of concern and a recent outburst of technology with this question, there are some indications that a positive answer to this generic question may be less forthcoming than suggested by a superficial review of the extraordinary new technological developments that make it possible for us to even consider asking this question. The literature of cognitive neuroscience grows each year as greater and greater investments in time and energy are made by the scientific community, particularly with regard to the macroneural approach. There seems to be an increasing feeling supported by new analytical and empirical findings that the macroneural approach, dependent as it is on the brain-imaging technology, is not going to provide the foundation for an overarching neuroreductionist theory of cognitive processes. Not only are the data unreliable, but also the analysis techniques are arbitrary and inconsistent.

The following list of general conclusions may help to set the stage for understanding the arguments leading to this disappointing negative conclusion:

1. There is increasing reason to support the argument that the mind–brain problem may be intractable at both the macroneural and microneuronal levels. Many of these reasons are based on conceptual or logical foundations, but many others are empirical—simply considering the number of neuronal elements involved in any mental process raises practical issues of computability that have no current solutions.
2. Theories come in several different kinds. Not all are neuroreductionist; some are purely descriptive. However, there is considerable discussion about what constitutes a neuroreductive theory or whether such an ideal understanding at any level is actually possible. Scientists still argue over the descriptive versus the reductive nature of theories.
3. Macroneural theories suffer from the loss of what may be critical data because of the pooling or summation of microneuronal signals.
4. Microneuronal theories, conversely, suffer from the very large amount of data that must be processed to answer the mind–brain question. Empirical testing of the theories may be impossible for the most practical reasons.
5. Obviously, science is driven and constrained by the available measuring instruments. It is not so obvious, however, that it is not only our empirical results but also our theories that are so governed. Technology can both enable and mislead explanations.

6. A main goal of macroneural theories is the localization of cognitive processes in specific regions of the brain. This may be a dead end.
7. A main goal of microneuronal theories is the study of neuronal networks of realistic cognitive size. This, too, may be a dead end.
8. Other factors that might mislead theoretical endeavors include:
   - Incorrect attribution to the wrong level of analysis
   - Poor definition of psychological constructs
   - Inaccessibility of cognitive processes
   - False analogies between transmission and integration processes

These summary points are presented here to clarify my prejudices as we set off on a more detailed discussion of microneuronal theory in cognitive psychology. Chapter 2, in accord with the general microneuronal goal of this book, deals with theories involving single neurons. Chapter 3 is an introduction to networks, the ways in which we study them, and the problems generated by their intrinsic complexity. Chapter 4 deals with relatively simple theories in which a few neurons are assumed to be interconnected. This intermediate level of neuronal networks is then followed by what is probably the frontier of modern neuroreductionist theory—the application of today's supercomputers and the problems that arise from their application to the potential solution of the mind–brain problem—in Chapter 5. Chapter 6 summarizes the argument put forward in this book.

Notes

1. In several places in this chapter I have abstracted material from my previous writing to review and clarify a necessary point.
2. Some of the material in this section has been adapted and updated from an article by Uttal (2015).
3. A readable technical discussion of the physics of fMRI recording can be found in Uttal (2001).
4. This trend may not be continuing; there is some suggestion that the number of papers that more or less naïvely attempt to localize particular cognitive functions in particular regions of the brain is slowing. My general impression is that some of our most prestigious journals (e.g., *Science*) are not publishing articles of the kind that contributed to this statistic. Indeed, even some of the specialty journals (e.g., *NeuroImage*) long associated with correlations of cognitive states and brain images seem to be more heavily populated with technological issues rather than simplistic assignments of cognitive processes to localized brain regions than previously has been the case.

2

SINGLE NEURON PRACTICE AND THEORY

2.1 Single Neurons

Before the discovery that the nervous system was made up of cells that were demarcated from one another and from their extracellular environment by a gossamer membrane composed of lipid molecules, the nervous system was thought to be a single syncytium with protoplasmic continuity among the neurons (an idea attributed to Joseph von Gerlach, 1820–1896). A new idea—that neurons were protoplasmically separated from one another by membranes—was first promulgated in 1891 by Wilhelm Waldmeyer (1836–1921). The new concept grew directly from the development of a new staining technique invented by Camillo Golgi (1843–1926). The key to the Golgi stain's importance was that only a small proportion of the neuron's bodies, axons, and dendrites were stained by this technique. Furthermore, staining seemed to stop abruptly at what could only be interpreted as membrane boundaries. The implications of the resulting microscopic images were championed by Santiago Ramón y Cajal (1852–1934) as reflecting a clear-cut separation of the intracellular protoplasmic material from the extracellular fluids. Although Golgi never accepted this new idea (which came to be called the Neuron Doctrine), both he and Cajal received the Nobel Prize for their respective contributions in 1906.

The initial idea behind the Neuron Doctrine was a purely anatomical one—namely that the cells of the brain are not protoplasmically interconnected. Instead, as a result of a diversity of recording and microscopic staining techniques, it became clear that what had been thought of as a reticulum of protoplasmically interconnected neurons was actually a network of neurons that were only functionally interconnected. Neurons, it became clear, maintained their individual identities and the boundaries between adjacent neurons even as they participated in the ensemble activities of the whole brain.

The enunciation of the Neuron Doctrine changed neurophysiological thinking in profound ways. Not only did this lead to new experimental approaches but also new modes of theorizing about mind–brain relations eventually emerged. Impelled by the Neuron Doctrine, a fundamental question immediately arose—how did these separate cells communicate with one another? The answer was provided by Charles S. Sherrington (1852–1952) when he named the main mechanism of interneuron communication—the synapse. Otto Lowei (1873–1961) uncovered the chemical nature of synaptic connectivity, and Henry H. Dale (1875–1968) specifically discovered the first chemical transmitter substance (acetylcholine). At first the synapse could only be defined functionally; it was so small that it could not be imaged until the development of the electron microscope in the 1930s by Ernst Ruska, a contribution for which he received the Nobel Prize in 1986. A full discussion of the history and nature of synaptic activity can be found in many books (e.g., Bennett, 2001; Pickel and Segal, 2013).

Many of the properties of the neuron are determined by the electrochemistry of the cell or bilipid plasma membrane. This membrane is the boundary between the interior of the neuron and the outside environment. It is made up mainly of two layers of a particular lipid molecule, the heads of which are attracted to water (hydrophilic) and the tails of which are repelled by water (hydrophobic). As a result of these attractive and repellant forces, membranes can spontaneously organize themselves in an environment that contains the necessary lipid molecules.[1] It also is a powerful force to maintaining the integrity of the membrane during mechanical insults—a factor that will become of central importance in our later discussion.

The important influence these new discoveries had was to emphasize the role of the individual neuron in both experimental protocols and theoretical thinking. During the first half of the 20th century, a considerable amount of attention was paid to determining the basic properties of neuroelectrical activity in terms of the selective permeability of the cell membrane to ions. During this time, neurophysiologists developed techniques for recording from single neurons. This work culminated in the Hodgkin and Huxley (1952) model of neuronal activity, a feat for which they also won the Nobel Prize in 1963.

Even more important to the development of theories of cognitive processes was the discovery that patterned stimuli, more so than their raw energy, determined the responses of individual neurons. The corollary idea that the patterned activity of individual neurons could encode or represent properties of the stimulus, at least in the peripheral nervous system, gained wide popularity in the mid-20th century. It was, however, a huge conceptual step to the assertion that single neuron activity could explain much more subtle and complex symbolic cognitive processes; in other words, that complex cognitive processes and events, not just the basic parameters of magnitude, quality, or timing, could be encoded by the activity of individual neurons.

The idea that a single neuron could also encode more complex concepts such as a particular face or object persists to this time. However, single neuron theories remain contentious in the context of today's network theories—theories that emphasize the alternative view, namely that the pattern of activity in broadly distributed, multi-neuron networks is the key to understanding cognitive information processing by the brain.

This chapter has two parts. First, it tells the story of those single neuron research technologies and their strengths and limitations in our search for understanding of the mind–brain problem. We start with an elaboration of the methods and techniques that are used to determine the functions of single neurons. Of one thing we can be sure—the availability of certain technological instruments (e.g., high-impedance microelectrodes) had a defining influence on the set of theories we consider in this chapter. The second part is an evaluation of the theories that are based on single neurons.

2.2 The Technological Bases of Single Neuron Theories

As just noted, neurons are now appreciated to be the basic cellular elements of the nervous system. Although they come in many shapes and sizes, neurons, like all other cells of the body, have evolved to carry out a specific function—they are specialized for information transmission and integration. That is, their role is to convey neuronal signals representing information from one part of the body to another and then to integrate this information so that coherent and adaptive responses can be made by the organism. The ultimate role of the nervous system is therefore to provide adaptive responses to environmental stimuli—a process that requires the gathering of sensory information, its integration into meaningful cognitive interpretations, and the selection of appropriate responses. Although neurons, like any other cell of the body, may perform other supportive, metabolic, or genetic functions, their main role is information processing.

Information, in the sense that I use it here, can be encoded in terms of patterns of organization of many neurons as well as the states of individual neurons. Conveying information means the transmission of signals from one part of the nervous system to another; integration refers to the mixing and mutual interaction of multiple signals into coherent patterns that can produce adaptive behavior in the form of both overt motor and covert mental activity. The currently most popular basic postulate of cognitive neuroscience is that "cognition" (or any other word you may choose to designate mental activity) is nothing more or less than the pattern of neuronal activity created by these transmission and integrative processes.

In the past half-century we have learned an enormous amount about how individual neurons function. We now know about the physics and electrochemistry of how they are stimulated, the kinds of chemical and electrical responses they produce, and how they communicate with one another synaptically. What we do not know very much about is how psychological processes arise from this biochemical and electrical activity. Indeed, hopefully without falling over a philosophical cliff, we might ask if psychological processes are necessary or just a meaningless epiphenomenal byproduct of neuronal activity. This may seem like an extraordinary question; however, considering that a fully functional brain could *in principle* function without any consciousness (as a "zombie brain" as discussed by Chalmers, 1996), the possibility that we are actually automata enjoying the unanticipated consequence of personal sentience as our nervous systems go about their business cannot be ignored. The basic problem is, as I have noted, that we do not yet know what the relationship is between neural activity and cognitive activity either in terms of single or multiple neurons.

Many neurons specialized for information transmission over long distances have evolved elongated processes (axons). For example, a single axon runs from the foot to the spinal cord before synapsing, thus providing a communication line for an irritation in the toe to be interpreted by the brain as a pain. The axon is an elongation of the neuron's cell body extending from the cell body or soma, as it otherwise known, to a distant neural locus where it may communicate via a multitude of synapses to the next neuron in an extensively interconnected network. Other types of neurons specialized for local information processing may have very short axons with many connections spread over its surface. In general, axons terminate in close proximity to the receiving region or dendrite of the next neuron.

Synapses are now known to be of two types, either chemical or electrical. Chemical synapses conduct by emitting specialized molecules known as "transmitter substances" from the "presynaptic" axonal region of the neuron. These molecules are received by specialized molecular receptor sites on the "postsynaptic" dendrites. Electrical synapses are rarer, more tightly packed than chemical synapses, and cannot amplify neuronal signals as do chemical synapses.

Because synapses are so small, their history is much more recent than are those of other neuronal components. Observation of synapses requires the extreme magnification provided by electron microscopes—a gain that must be as high as 2 million compared with the maximum magnification of an optical microscope, which is limited by the diffraction of light to about 2,000. In the past couple of decades, ingenious fluorescence techniques (e.g., Hell and Wichmann, 1994) suggest that new optical technologies may permit "superresolution" beyond what is imposed by diffraction limits on optical microscopes.

To accelerate the conduction time, elongated axons are usually sheathed in a multi-layered myelin coating composed of a series of specialized and overlapping Schwann cells. This myelin sheath is discontinuous with narrow constrictions referred to as "Nodes of Ranvier" spaced periodically along the length of the axon. The electrical signals that produce the propagated neural response along an axon are able to jump from node to node, producing a much higher conduction speed than would be possible with an unmyelinated fiber. Cells primarily designed for localized interaction have much shorter axons that are typically not myelinated—indeed, some presynaptic axons are so short that they are hard to distinguish from the dendrites of the receiving or postsynaptic neuron. The cell body, or soma, is a complex structure with many components and metabolically active microstructures, including the intracellular mitochondria, the synaptic structures, and the capillaries that convey blood to and from the neuron. It is also important to note that most neurons are surrounded by supportive glial cells that have long been thought by some (e.g., Galambos, 2007) to have some role in information processing beyond the purely supportive role usually associated with them.

Penetrating the membrane are also some relatively large molecules mainly concerned with the transport of metabolically significant chemicals from the outside to the inside of the membrane and vice versa. This flow of materials includes both simple ions, whose ebb and flow are dependent on the transmembrane potential, and synaptic transmitter substance of complex molecular configurations. It is now thought that large penetrating molecules of this kind are both the presynaptic source of both transmitter molecules and the postsynaptic gates that permit the transmitter substance to enter a neuron. It is the intricate pattern of interactions among neurons based on these robust molecular engines that account for the neuronal activity that in some mysterious way produces all of our cognitive processes.

The next question with which we must deal is how we go about observing the activity of neurons, synapses, and membranes at the microneuronal level at which they operate. Although much has been accomplished in understanding their biochemistry and anatomy, it is at the interface between neurons and electronic equipment from which most of our information about their interneuronal information processing functions has come. The method of stimulating and recording from single neurons is a highly developed technology that once again illustrates the close connection between our technologies and our theories.

Electrodes are the point of contact between our measuring instruments and the activities of neurons and the brain. It is here the biology and electronics interface and the micro- and macro-events of the nervous system become the stuff of empirical science. Over the years, there has been a steady progression of more and more sensitive instruments that permit recording from and sending stimuli to the nervous system. Electrode design has become a highly developed technology unto itself with different specialized electrodes serving different needs. Somewhat arbitrarily I have distinguished between stimulating and recording electrodes although in many cases the same kind of electrode may serve either function.

Recording Electrodes

The basic role of recording electrodes is to detect and measure the electrical activities produced by the various components of the nervous system.[2] Electrodes come in a wide variety depending on the specific application. Different electrodes are necessary for measuring the compound (i.e., pooled) neural responses than for measuring the isolated responses of single neurons; electrodes that are optimal for intraneuronal responses may be quite different than those that are useful for extraneural ones. Other electrodes are specialized for recording the macroneural responses such as those picked up by the electroencephalogram (Berger, 1929) or whole nerve preparations (Dawson and Scott, 1949); these signals may require only simple electrical contacts—a stainless steel button taped to the skull or attached to the arm does a satisfactory job of communicating these macroneural signals to our measuring equipment. In other contexts, there has been a persistent effort over the years to make direct contact with brain tissue by means of surgical techniques that use implanted electrodes. (A comprehensive discussion of electrode technology can be found in an informative article by Cogan, 2008.)

Our main interest in this section is concerned with the electrode types that can be used to measure single neuron responses. (In Chapter 3, I will deal with multiple arrays of electrodes that can be used to record from many neurons simultaneously.)

Neurophysiological recording of the electrical activity of neurons is accomplished by measuring the minute voltages produced by shifting ionic distributions across the cell membrane. To do this, some kind of an electrically sensitive contact—an electrode—must be placed in or near the neural tissue. Subsequent amplification of the obtained biological signals to a level that is sufficient to drive recording equipment is accomplished by appropriate electronic devices. The amplified signals can then be connected to a display such as a cathode ray oscilloscope or entered into a computer for subsequent analysis.

The critical and most idiosyncratic component in such a system is the initial contact point—the actual electrode interface that senses the tiny neuronal electropotentials and conducts the information they carry to signal processing equipment. Various forms of electrodes capable of observing single neuron responses have been used in the history of neuro-electrophysiology depending on the particular neuroelectrical signal of interest. As we shall shortly see, electrode technology has driven the course of modern neurophysiology and with it the theories that have been engendered.

The first attempts to record electrical signals from single neurons were carried out on surgically excised whole nerves consisting of a bundle of axonal fibers. The electrodes used to record these signals were typically salt-saturated fiber wicks. The experimental design was a direct one—whole nerves were surgically dissected free of their surrounding tissue and laid across the wick; the detected voltage responses were connected to an amplifier and recorded on whatever apparatus was available for displaying and recording. Such techniques, however, produced signals that were the accumulation of the responses of many neurons, a mélange that may have hidden the activity of the individual neuron, as shown in Figure 3.2.

A step toward single fiber recordings occurred when investigators were able to dissect single axonal fibers free from a compound nerve. Wick electrodes were also at first used to detect the electrical activity generated by these single dissected axons. However, the extracellular responses so recorded were very small (a few microvolts) compared with the actual transmembrane resting voltage difference that was ultimately measured across the membrane; these transmembrane voltages could be as large as 70 or 80 millivolts or even larger in some cases.

The next important step in measuring neuronal activity occurred when Young (1936) discovered that the squid *Loligo vulgaris* had evolved a neuron with a very large axon—one that could be as large as one millimeter in diameter. This was large enough to permit a reasonably sized glass tube electrode filled with a conducting electrolyte solution to be inserted into the cut end of this giant axon and, thus, into its internal environment. The glass tube, therefore, served as an intracellular electrode. For the first time, using the accidental availability of the anatomical anomaly in this model preparation, it became possible to observe the transmembrane voltage in its full splendor, not just the much diminished version obtained with extracellular electrodes. When a reference electrode (all voltages are measured as potential differences between two points) was placed in the extracellular fluid, the full transmembrane voltage could be measured as the voltage between the inside and the outside of the neuron.

The enabling "trick" in this case was getting one relatively large electrode into the interior of an unusually large neuron. Thus, the accident of an evolutionary freak—an unusually large axon—made it possible to develop the ionic theory of neuronal action. Using the giant neuron of the invertebrate squid *L. vulgaris* as a model of vertebrate neuronal ionic mechanisms, Hodgkin and his colleagues (Hodgkin and Katz, 1949; Hodgkin and Huxley, 1952; Hodgkin, Huxley, and Katz, 1952) were able to work out many of the details of the ionic flow across the membrane that accounted for its ability to both create local potentials and to rapidly propagate a signal from one end of the axon to the other—research that earned them the Nobel Prize in 1963. It also permitted them to develop a comprehensive theory relating membrane potentials and ionic concentrations that persists virtually unchanged up to the present time.

Stimulating Electrodes

Electrodes have to be more than just one-way devices. In addition to their important uses as sensitive transducers of neural signals, they can, as noted previously, also be used to stimulate or activate neural tissue. For macroneural stimulation, such as is necessary to evaluate percutaneously the properties of peripheral nerves (e.g., Uttal, 1959), a simple, hemispherical, stainless steel button, buttressed by some electrical gel, is all that is required.

Similarly, when stimulation of the brain with large implanted electrodes was required, the material from which the electrodes were made was relatively unimportant—almost any nontoxic and corrosion-resistant piece of wire would be satisfactory. In some situations (e.g., Uttal and Krissoff, 1966), in which a diffuse spatial stimulus was required, it was sufficient to simply insert two fingers into separate test tubes filled with saline solution.

The Modern Microelectrode

The anatomical freak typified by the size of the squid's giant axon, however, is not available in vertebrate neuron preparations. Vertebrates, despite their larger size and greater number, generally have much smaller neurons than do invertebrates. To study their individual nature required an alternative to pushing a rather large glass pipette into the cut end of a fortuitously supersized axon. It was at this point that the bilipid molecular layer of the membrane played a critical role in continuing progress toward a full understanding of how neurons operate, especially how the action potential—the propagating "spike" of electrical activity accompanying the ionic flows—communicated information from one part of the nervous system to another.

The attractive hydrophilic forces between the lipid molecules that made up the cell membrane were sufficiently strong that it turned out to be possible to punch an appropriately small (of the order of a micrometer) electrode right through the cell membrane without immediately destroying it. The intermolecular forces of the lipid molecules were strong enough to provide a robust seal around the electrodes—a seal that could last for many hours.

Among the earliest to develop such a membrane-penetrating microelectrode were Ling and Gerard (1949) in their pioneering study of the resting potential of frog muscle cells. The technique had been under development for a number of years in Gerard's laboratory by Graham (Graham and Gerard, 1946), but was finally brought to a mature reliable stage by Ling.[3]

The technique that Ling and Gerard developed to manufacture a microelectrode required that a glass pipette be heated at its middle until the glass began to melt and then simply pulled from both ends. This procedure reduces both the external diameter of the pipette and its internal diameter of the interior down to approximately 1 μm in diameter—a size comparable to the diameter of the neurons under study. Indeed it provided two electrodes as the "pulled" glass pipette was carefully broken apart at the hot spot. The ability to produce such a miniature electrode—a microelectrode—lies in the fact that the geometry of a hollow glass pipette (like those of the millefloria of Venetian glass) is preserved even as it is vastly demagnified in diameter by the heating and pulling. The important part of this procedure is to preserve the tiny (as small as a micron) hole at the end of the glass tube so that the conducting fluid it contains can make electrical contact with the intracellular fluid when it is punched through the cell membrane of a neuron. The final role of glass in this manipulation is simply to act as an insulator between the intracellular and extracellular fluid, thus avoiding a short circuit across the membrane.

The miniaturized glass microelectrode is then filled with an electrolyte such as Ringer's solution—a mixture of any of several salts, potassium acetate, or potassium chloride. Many other solutions can be used as the electrical interface as long as they are conductive and do not disrupt the metabolism of the neuron being studied by the unintentional injection of chemicals that may be toxic. The other, un-demagnified end of the glass tube (from which the electrode was pulled) remained large enough for a conducting metal wire to be inserted to serve as a reference electrode. The functional role of this intracellular microelectrode is to convey the relatively large transmembrane potentials to a preamplifier; from there, signals could be conveyed to display and computer analysis devices.

An early application of the new glass pipette–microelectrode technology to neurons from the frog's sciatic nerve was reported shortly after Ling and Gerard's publication. The first application of a microelectrode inserted into the human brain was reported by Ward and Thomas (1955).

An alternative method of making intracellular micron-size microelectrodes is to electrolytically polish a metal (usually tungsten, platinum, or iridium since steel points are too brittle), wire down to a 1-μm point, and then coat the wire with some kind of insulating varnish (Grundfest, Sengstaken, Oettinger, and Gurry, 1950; Hubel, 1957). The tip is left uncoated so that electrical contact can be made with the neuron, usually extracellularly. Both forms of electrode—glass micropipettes and sharpened, then insulated, metal points—thus consist of a tiny conducting core surrounded by insulation (glass or varnish) that can be used as a membrane-penetrating, active electrode. The second point of electrical contact, the reference electrode, can be placed virtually anywhere in the extracellular fluid environment of the preparation and is decidedly low-tech—a clip or plug can be attached to any available tissue in the specimen whose nervous system is being studied.

The glass microelectrode is particularly useful for intracellular recording. It is capable of sensing a maximum amplitude transmembrane voltage from a single neuron. Insulated electrodes made from many different kinds of metals seem to be better adapted to extracellular recording, typically resulting in a mixture of neuronal signals from several nearby neurons. How many neurons will be sensed is determined by fortuitous accidents of the arrangement and propinquity of the neurons and the electrode and is typically not under the control of the experimenter.

In sum, microelectrode design is still very much an intuitive art form, each investigator often constructing their own design to meet the special needs of his or her experiment. Even the original work reported by Ling and Gerard was the result of a multi-year effort by several people before their design was stable and reliable enough to be used in any quantitative manner. For reasons that are still not entirely clear, minor details in the electrical design or in the material from which a microelectrode is constructed may have significant effects on the quality of the signals they detect.

Recording Devices

One of the earliest devices designed to store and display the electrical activity of the nervous system was Willem Eintoven's (1860–1927) invention—the string galvanometer. Although first used to record the electrocardiogram, eventually it was applied to the measurement of the conduction speeds of nervous impulses in situations in which its predecessor—the capillary galvanometer (in which the height of a column of mercury was used as the indication of electrical activity magnitude)—was insufficiently sensitive. In either case, displacement of the string or the meniscus of the mercury column by neuronal response–generated electrostatic forces was recorded photographically. These primitive displays of electrical response activity have been completely replaced by the modern cathode ray oscilloscope, an evolutionary outcome of the Crookes tube, first invented by William Crookes (1832–1919). In the Crookes tube, a stream of electrons inside an evacuated glass tube is directed onto a fluorescent surface. This directional control is affected by the application of external magnetic fields. With such modern equipment, this basic idea of changing the position of a spot on the face of an oscilloscope as a function of the applied voltage makes it possible to plot the time course of the electrical waveforms generated by neural activity.

It is something of a historical curiosity that the cathode ray oscilloscope was not used in cellular neurophysiology until 1922 when it was first applied by Joseph Erlanger (1875–1965) and Herbert Gasser (1888–1963) despite its invention as early as 1897 by Karl Ferdinand Braun (1850–1918). This was likely due to inadequate amplification of the tiny signals generated by neurons, rather than any fundamental limitation on the part of the oscilloscope itself.

An especially significant development for studying microneural events, therefore, was the emergence of modern high-input impedance preamplifiers to extend the range of the cathode ray oscilloscope to low voltage levels. The small size of the electrical potentials recorded from the nervous system meant that they had to be significantly amplified before they could be reliably measured. The invention of the triode vacuum tube amplifier by Lee De Forest (1873–1961) was a milestone not only for amplifying music but also for neurophysiology. The triode vacuum tube was constructed from an anode and a cathode and a control electrode, all of which were encased in an evacuated glass enclosure. The amplification of small signals was accomplished first by the three-electrode vacuum tube and then by modern solid-state circuit designs. By applying the preamplified, but still relatively small, signal generated by the neuron to the control electrode situated between the cathode—a source of electrons—and the anode—a sink for electrons—small signals could be amplified to a level that was adequate to drive the position of the electron beam. The basic principle of this kind of amplification was accomplished because the small signal placed on the control electrode could regulate a much larger current flowing between the cathode and the anode. Modern amplifiers, of course, are now constructed from solid-state components, but perform the same amplification function albeit at ever-improved levels of amplification, noise reduction, and frequency response.

Amplifiers and Impedances

Unfortunately, in meeting the need for the ultraminiaturized size of an intracellular microelectrode so that the neuron would remain reasonably intact when impaled, another problem is created. Both the tips of glass and the metallic microelectrodes are so small that they present very high electrical impedance (resistive impendences of microelectrodes may be as high as 100 megohms); this requires special electronic preamplifiers or head stages prior to the main amplifier for signals to be measurable.

There are two problems with high-impedance electrodes. First, their resistive impedance (R_m) is part of a voltage divider circuit with the input impedance of the preamplifier R_p as shown in Figure 2.1. The voltage that is detected by the preamplifier, therefore, is a function of the impedance of the electrode and that of the amplifier in accord with Equation 2.1:

$$V_p = V_m \times R_m/R_m + R_p \qquad \text{Eqn. 2.1}$$

where V_p is the voltage sensed by the preamplifier and V_m is the voltage produced by the neuron. Thus, if R_m is much larger than R_p, then V_p (the portion of the voltage) across the amplifier input will be very small—perhaps undetectably so

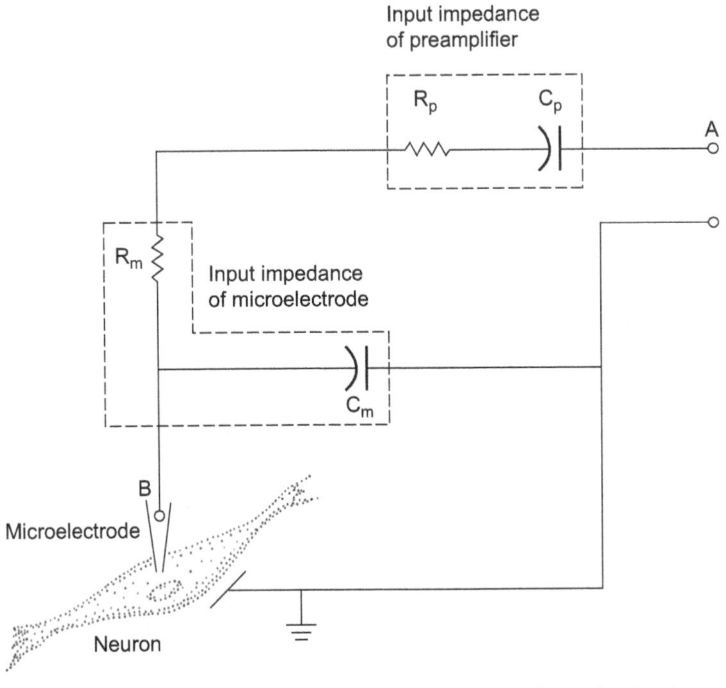

FIGURE 2.1 The electrical impedances involved in a microelectrode circuit.

Reproduced from Uttal (1975).

by the main amplifier. The net effect of this will be to make the tiny voltage from the neuron virtually undetectable by the amplifier unless the amplifier input impedance is itself high. This is the reason that specialized high-input impedance preamplifiers are so necessary in cellular neurophysiological research.

The second main problem with high-impedance microelectrodes is that they operate in combination with the capacitance of the system to limit its temporal response properties in accord with the RC rule:

$$TC = R \times C$$

where TC is the time constant of the system, R is its resistance, and C is its capacitance. The TC of such a circuit will determine the overall frequency response of the recording system; the larger the TC, the less the high-frequency components of the signal will pass through the system. The required high-frequency response of a neuron-preamplifier combination must be measured at least in kHz for neuronal responses to be detected and their waveform reproduced although some modern experiments may require much higher limits.

Impedance mismatches can also introduce spurious noise of several different kinds. The engineering of preamplifiers with sufficiently high-impedance properties to match microelectrodes is an art unto itself, but one that has matured to such a high level of development that they are currently available off the shelf from a number of different manufacturers.

Patch and Voltage Clamps

Other instrumentation tools that have become mainstays of single neuron physiological reseach are the voltage clamp and the patch clamp. The problem arising is that the several ion distributions that collectively sum to determine a neuron's membrane potential are each dependent on the membrane potential itself. Because this is a dynamic processs with an intrinsic form of feedback, it is diffiult to determine such factors as the specific relationship between the membrane potential and the flow of the particular ions that collectively make up the actual neuronal response.

The voltage clamp overcomes this difficulty by generating electronic feedback signals to stabilize the membrane potential at a predetermined constant level. This essentially locks the membrane potential at any prespecified level. The neurophysiological investigator is then able to derive the relationship between stable membrane potentials and the ionic distributions. By varying the feedback signal, the entire set of functions relating the several kinds of ion flow and membrane potentials across the membrane can be determined.

The patch clamp is a derivative of the voltage clamp idea but is designed to work only a small patch of the cell mebrane. This patch is often so small it is possible to isolate a single one of the molecular channels through which ions flow during neuronal membrane activity. Attachment to such a small region is accomplished by attaching the membrane to the tip of an electrode by means of a vaccum. Both voltage and patch clamps have made it possible for modern neuroscientists to understand how the flows of the various ions (mainly Na+, K+, and Cl-) are dependent on the electrical properties of the membrane.

However, this detailed explanation of the ionic basis of neuronal activity does not speak directly to the problem at hand—how cognition and mind are produced by this elaborated ionic dance. Putative explanations based on single neurons are discussed in the next section, in which I review how some theories of cognition attempt to link cognitive processes to individual neuronal responses.

2.3 The Single Neuron Theoretical Approach to Solving the Mind–Brain Problem

In our search for the answer to the great mind–brain question, we seek to find the neurophysiological properties and activities that "account for," "encode," or "represent" our cognitive experiences. All of these words are ambiguous and ill-defined, but the general idea is that there is in the brain some corresponding neurophysiological activity that is not just correlated with mental activity but "is" mental activity. In other words, mind and all of its accouterments are assumed to be, in principle, identifiable in some way with the physiological and anatomical properties and activities of neurons—either single neurons or networks of them and possibly of other cell types in the brain. A better term to explain what is meant by this relationship is the *psychoneural equivalent*—the neurophysiological response that would be the necessary equivalent to cognitive experience. What the psychoneural equivalent of a cognition is, of course, is only hinted at in current cognitive neuroscience research but, under the influence of the methodologies and technologies described in this book, speculative theories have proliferated.

As I begin this discussion, therefore, it is useful to be very specific about what is meant by single neuron theory as well as its antithesis—neuronal network theory. Single neuron theory (in its most extreme form) in cognitive neuroscience refers to a particular set of possible explanations of the neural bases of cognitive processes that are expressed in terms of the anatomy and physiology of just one neuron (or, according to Barlow, 1972, just a few). According to the most fundamental version of single neuron theory, the activation of a particular neuron will elicit the experience of some object, person, or even a very complex intangible concept such as "freedom."

Once again, this illustrates the enormous influence that our technological tools played in our theoretical development. It is all but certain that, without convincing empirical support for the Neuron Doctrine and abundant single neuron findings, the idea of individual neurons as the conveyers of our psychophysical experience would have never occurred to anyone. Instead, more popular would have been the holistic idea in which all or most of the brain is involved in mental activity. (Holism, of course, is also driven by technology, namely the EEG and fMRI.)

There are several basic problems inherent in single neuron theoretical formulation that suggest that it is not the ideal route for theorizing. First, we do not know how many "concepts" there are in our mental life and so we cannot tell if even the large number of neurons in the brain is adequate to represent all of them, each with its own neuron whose activities serve as the psychoneural equivalent of some mental experience. Counting the number of available neurons and associating each with its cognitive equivalent, although possible in principle, probably would not be a practical approach.

Second, unlike the first problem, which could in principle, if not in practice, be validated by simply finding neurons whose response could be correlated with a concept, there is no way that we could carry out the necessary experiments to validate the association between specific cognitions and single neuronal activities. Although it is reasonable to assume that we could find a neuron that responded to a particular stimulus, locating a particular neuron whose activation provided empirical proof of a particular subjective awareness would be, to say the least, a very challenging task.

Third, practical issues such as the manipulation of a particular neuron (isolating it, turning it off and on) would always conflate, confuse, and confound the results of any experiment attempting to show single neuron representations of cognitive processes.

An important related issue is—how do we account for the selective sensitivities and responsiveness of individual neurons to very complex "concepts"? In general, the basic anatomical mechanism purported to explain these findings is a progressive convergence from the properties of a stimulus onto fewer and fewer neurons, with the result that more and more complex ideas or concepts can be represented by fewer and fewer neurons. Ultimately, this convergence approach would respond selectively to complex gnostic concepts and entities such as "yellow Volkswagens" or "grandmothers" or "altruism." This extends the fundamental idea of a hierarchy of increasingly complex responses in the visual system originally suggested by Hubel and Wiesel's (1968) progression of simple, complex, and hypercomplex cells. The concept of a convergent hierarchy suggested that the lower, simpler, spatial-temporal properties defined at the stimulus level (such as line orientation) might progressively converge with other properties onto higher-level concepts, eventually as specific as a particular human face, at the same time that neurons are becoming more and more selective and particular to fewer and fewer tangible ideas.

The idea of a progressive convergence of constituent codes onto neurons suggested, to hammer in this point, that the meaning represented by higher-level neurons becomes more and more specific until complex cognitions such as particular objects, people, or concepts are represented by the activity of a single neuron. The important mechanism underlying further development of this idea is, thus, a progressive convergence of the tuning of a neuron from the spatial-temporal domain of simple stimuli to the gnostic, conceptual, or symbolic domain.

At present, however, no one has been able to demonstrate this progressive neurophysiological convergence once one proceeds beyond the peripheral sensory systems.

In sum, the core of single neuron theory is that once it has been established that a neuron has a particular sensitivity (such as responding to an individual face) subsequent activation would be the necessary "cause" or "equivalent" of the mental experience of that face. The basis of a pure single neuron theory, therefore, is that the activation of a single neuron *is* the psychoneural equivalent of entities of unlimited complexity.

Without a doubt, a statement that asserts that single neuronal activation is tantamount to experience is replete with uncertainties, ambiguities, and unresolved subissues. What are the other billion neurons doing while the critical one is determining what we momentarily perceive or think? Is there some kind of priority that one neuron has over all others? How did the neuron acquire the particular sensitivity? Is the individual neuron itself conscious or its response sufficient to determine a unique conscious experience? Is the role of psychoneural equivalent being conflated with its equally plausible role as a "trigger"? For example, if activation of a single neuron is simply acting as a trigger for a more complex response actually embodied by a network of multiple neurons, then it might well appear to an experimenter to be causal, but any such attribution would be fallacious. In short, the single neuron theory, although superficially simple, is actually extremely subtle and complicated by questions rarely asked and almost never answered.

The antithesis to the single neuron theory is the neuronal network approach. Rather than ascribing cognition to the activity of a single neuron, this alternative proposes that the psychoneural equivalent is embodied in the interactions of a network of many neurons, all of which are interacting with each other in complex and dynamic manners. Controversies in this case concern how many neurons are involved in a network and whether we can quantify network activity in a mathematically tractable form. The raw conceptual simplicity of the single neuron approach (neuronal response X is the equivalent experience Y) is an advantage not shared with neuronal network theory. Even at the earliest stages of prototheory development describing how brain activity might make the transition from neural function to psychological experience, preliminary anatomical evidence made it clear that the process would be an overwhelming one that was not likely to be solved in the foreseeable future if phrased in terms of the network interactions.

Both single neuron and network theories propose an answer to the question—what activities and properties of the brain produce, cause, or become cognition? Both, however, raise many issues of meaning and interpretation that have to be considered in detail. In subsequent chapters, I consider neural network theories in detail; here I now discuss the practical and conceptual implications of single neuron theories. However, before going on, there is one important clarifying comment to be made—both the single cell and the neuronal network theories have a common feature. That is, both maintain information about the status and activities of individual neurons. Thus, both fall into the microneuronal category. Although this may not seem to be the case when one observes the massive neuronal networks that characterize the brain, this is their explanatory level. The differences with the macroneural approach are profound.

Single Neuron Theories

The earliest single neuron theories antedated the neurophysiological findings that were eventually to provide the putative evidence for what remains a popular theoretical approach. William James, in his classical 1890 psychology text, spoke of the "consciousness" of *each* and *every* "brain-cell." To Charles Sherrington (1940) is attributed the first use of the term "pontifical neuron," a phrase suggesting that an individual neuron might by itself represent a complete cognitive experience in some way, dominating the current cognitive experience.

More complete and up-to-date versions of single neuron theories were presented by Barlow (1961) and Konorski (1967) in the 1960s. In his version, Konorski distinguished between transit (sensory and motor codes) and gnostic (mental or cognitive representations) as distinguishable categories. This is an important distinction that is all too often lost in evaluating today's theories, since great success in understanding peripheral transmission coding has been achieved, whereas almost no compelling progress has been obtained in determining the neuronal equivalents of complex gnostic or cognitive experiences. Barlow's approach was somewhat more eclectic, evolving from a single neuron approach to one involving "a small multiplicity" of neurons—cells that he suggested behaved as "cardinals" rather than "popes" (Barlow, 1972).

The single neuron approach to theory building was further stimulated in the 1950s and 1960s by the extraordinary neurophysiological discoveries of such pioneers as Hartline and Ratliff (1957), Hubel and Wiesel (1957), and Lettvin, Maturana, McCulloch, and Pitts (1959), among many others. All of this research made the same point: response patterns could be observed in the responses of single neurons that seemed to parallel much more complex behavioral responses. Because many perceptual phenomena, such as visual illusions, had long been described phenomenologically but without any neuroreductive theories, analogies drawn between the associated peripheral transmission codes and the perceptual responses had a certain face validity, as well as a seductive attractiveness. Based on analogies between neuronal and psychophysical responses (and the success in defining the peripheral transmission codes), available single neuron neurophysiological findings quickly filled the vacuum as putative explanations of higher-level cognitive processes. In the decades following the 1950s, based on these pioneering studies of single neuron correlates of sensory information, the relatively simple idea that a single neuron could be responsible for representing complex perceptual experiences became extremely popular, if not dominant, in mid-20th century thinking.

Evidence provided support for the idea that cortical-level neurons become selectively responsive to particular human beings (e.g., the well-known, but hypothetical, "grandmother" neuron originally proposed by Lettvin, as cited by Gross, 2002). Although some controversy exists about the robustness of these findings (see the work of Tanaka, 1993, who showed specialization to face-like geometrical forms rather than particular faces), there does appear to be a relatively strong body of empirical evidence that suggests that these observations should be taken very seriously—specifically, the not-to-be-overlooked idea that there are neurons in the brain that respond only when very specific and very complex stimulus conditions are met. These highly specialized neurons, as I suggested earlier, are the end product of a process of progressive convergence that starts with sensitivity to the physical dimensions of a stimulus (e.g., its temporal-spatial properties) and progresses to neurons whose responses are related to the highly particularized meaning of a stimulus (e.g., a specific object, person, or idea).

A summary of a broad swath of current single neuron theories has been provided in an important review of the field by Spillmann and Werner (1990). In it, a distinguished group of investigators described their efforts to link the neuronal properties of the sensory pathways with perceptual experience—that is, to assume that these neurophysiological responses are the psychoneural equivalents of cognitive experiences. However, even then, there were voices calling out for a more thoughtful interpretation of the relation between these peripheral neuronal codes and perceptual experience. For example, Westheimer (1990) argued that:

> Once the first discoveries were made . . . single cell researchers in the mammalian visual system developed categorizations which do little for psychophysicists.
>
> *(p. 7)*

In retrospect, most if not all of the theories presented in Spillmann and Werner's book actually turned out to be associations of peripheral sensory transmission codes that conveyed information but did not themselves encode experience. The distinction being made, once again, is between transmission and experiential representation codes. To put the distinction simply, we should ask basic questions such as could the eye "perceive without a brain"? or could the occipital region of the brain "perceive without an eye"? or could any other isolated region of the brain "perceive" if activated in some indirect fashion? Unfortunately, empirical answers to questions of this kind are currently impossible to obtain.

The problem of how far up the sensory pathway one must go before we find neural mechanisms that are the equivalent of cognitive experience (rather than just communication codes) is rarely raised by cognitive neuroscientists. An exception is the work of Crick and Koch (1995), who asked the question, "Are we aware of neural activity in the visual cortex?" (p. 121). They suggested that there is no awareness of neuronal activity in visual area V1 but rather, that V1 is actually better considered as a part of the peripheral transmission system. If their suggestion is correct, then the many conclusions of cognitive encoding by single neurons based on correlations between psychophysical and neurophysiological responses in V1 may have been seriously misinterpreted. The correlated neuronal responses would actually be reflecting transmission codes, even as high as V1. Thus, the idea that neuronal responses in V1 were the psychoneural equivalents of our perceptions would be based on a false analogy—isomorphism of the dimensions of the stimulus and the psychological responses. This issue is especially important in the present context because any support of the cognitive efficacy of a single neuron's responses in V1 would become meaningless. Indeed, since the preponderance of the empirical work supporting single neuron theories has been carried out in the peripheral systems, most of it does not actually deal with the neuronal coding of awareness or experience per se.

Psychoneural equivalent codes at the level of cognitive activity are the answers to a different question than the one concerning transmission codes. The equivalence question asks—how do the neurophysiological responses become complex perceptual phenomena? If Crick and Koch are correct, although there may be correlations between neuronal activity and cognitive experiences, the relationship is not a causal one. That is, the neuronal responses are not the psychoneural equivalents of the experience but are conflations—both being driven by the common stimulus, but without any direct relationship between the neuronal and the psychophysical.

Without question, great progress has been made in the sensory domain and, for all practical purposes, most codes for the key sensory dimensions (i.e., quality, quantity, spatial extent, and temporal pattern) of most sensory stimuli in a variety of animals have been broken (Uttal, 1973). If explicating the transmission codes can be considered to be equivalent to theory development, then this is the most compelling argument that successful neuroreductionist single neuron theories have been forthcoming. However, this is not the same thing as demonstrating a solid link between individual neurons and higher levels of experience. Far less has been accomplished in understanding the relationships between complex cognitive processes and their underlying neurophysiological equivalents. Furthermore, if Crick and Koch are correct and transmission mechanisms (as opposed to cognitive ones) extend much farther up the ascending pathways into the visual areas of the brain than has been appreciated, then the gap between neuronal and cognitive responses remains as broad nowadays as it has been for decades.

In the following discussion, I briefly review some of the theoretical associations that have been made between single neurons and cognitive processes. I hope that my readers will keep in mind the admonition that there are no direct and robust links between neuronal responses and cognitive processes; rather, most of these "theories" are little more than associations by analogy or metaphors.

1. Intraneuronal lateral inhibitions observed in the Limulus visual system (Hartline and Graham, 1932) and in the frog by Barlow (1953) have been associated with visual phenomena such as the Mach Band (i.e., edge enhancement), the dark spots at the intersection of the Hermann Grid illusion, and a spatially broader perception of simultaneous contrast.

2. Lettvin et al. (1959) and Hubel and Wiesel (1965) observed that single neurons were selectively sensitive to natural stimuli such as moving edge detectors (flies?) and stimulus diminishment (looming predators?). The neurophysiological sensitivities were then associated with practical ecological needs of the entire organism such as "bug detection."

3. Enroth-Cugell and Robson (1966) showed that individual neurons displayed a differential sensitivity to spatial frequency—in particular the empirical result called the perceptual "tuning" curve. This was extrapolated to the idea that the Fourier frequency was associated with psychophysical detectability of spatial frequency patterns.

4. One of the most interesting single neuron theories of the 1970s suggested a near identity of neuronal and psychophysical responses. Jung and Spillmann (1970) introduced the concept of a Perceptive Field (PF), the psychophysical interaction between a spot stimulus and its surround. This perceptual phenomenon was deemed to be analogous to the Receptive Field (RF) of a single neuron in a neurophysiological experiment. Jung and Spillmann suggested that comparable perceptual results to the center-surround organization of a microscopic visual neuron could be obtained at the molar level. Psychophysical responses were measured to map out the surround of an isolated point stimulus. They argued that the dimensions of the PF corresponded closely enough to the dimensions of the RF of a single neuron to be the psychoneural equivalent of the PF.

There are, however, many differences between a single neuron's response and an analogous psychophysical phenomenon. One obvious functional difference is the neurophysiologically measured output of the results from an individual neuron, whereas the psychophysical experience must be the accumulated and overlapping result of many neurons. A priori, therefore, there seems to be no obvious reason why the two phenomena—the RF and the PF—should be of the same scale.

An obvious test of the PF hypothesis, therefore, is to ask—are the RF and the PF at least the same size or, to the contrary, are any similarities between them simply accidental responses occurring at different scales? Surprisingly, according to Spillmann (2014) the psychophysical PF and the neuronal RF are almost identical in size, a result obtained by comparing neurophysiological data reported by DeMonasterio and Gouras (1975) and Spillmann's psychophysical data.

Neri and Levi (2006) provided further support for the similarity of RF and PF sizes by reviewing seven pairs of experiments—one of each pair being an RF neurophysiological experiment and the other being a psychophysical PF experiment. They found several similarities between the two classes of experiments beyond simple size. Although this finding seems to be counterintuitive, the important point in the present context is that the RF of a single neuron is presented by theorists of this persuasion as a direct cause of the psychophysical phenomenon they call PF. However, even Neri and Levi point out that:

> The similarity between receptive and perceptive fields highlighted here should not be taken to imply that behavioral performance in these experiments was completely determined by the response properties of individual neurons in early cortex.
>
> *(p. 3472)*

5. Single neuron theories of sensory processing have also been proposed in the somatosensory system. The typical experimental paradigm in this case involves stimulating the skin mechanically, probing a single neuron in a peripheral nerve (e.g., the median nerve) with a microelectrode, and simultaneously asking subjects to report when they sensed the stimulus (see Valbo and Hagbarth, 1968, for an early example). Typical findings showed that the threshold of a subject was so low that a single action potential in a single nerve fiber could be reliably detected as a stimulus to the skin.

The inference drawn from this finding was that a single spike action potential was capable of encoding a complex experience such as cutaneous touch or pain. However, this inference is almost certainly misguided. The psychoneural equivalent for the experience is not likely to be the single spike action potential recorded from the peripheral nerve, but the greatly magnified burst of neural activity that is triggered by that action potential at higher levels of the ascending (i.e., central) nervous system. Thus, any attempt to use this as an argument for a single neuron theory of perceptual experience is fragile. The fact that our sensory thresholds are as low as they are is not evidence that single neurons are encoding experience. A more reasonable inference is that there is an enormous magnification in the neural activity; an increase that ultimately rises to the level of perception and ultimately to the relatively large amount of energy expressed in the verbal motor response—"yes, I see [or feel or hear or taste or smell] it." Indeed, macroscopic evoked brain potentials, EEGs, and brain images of all kinds show that vast numbers of neurons are generated in the regions of the brain associated with barely threshold stimuli. That a minute stimulus activating a single neuron might trigger a complex behavior is not evidence that the experience is encoded by that neuron; in fact, it seems more supportive of the role of huge numbers of interacting neurons in a massive network being involved in virtually any cognitive processes.

6. In the past few decades, a rush of reports has suggested that single neurons do respond to a very specific and important kind of stimulus—individual faces. No longer is the discussion dominated by the raw spatial or temporal properties of a stimulus such as moving edges or corners. Instead, neurons are being reported in various parts of the brain that respond selectively to faces. This may be in general terms (responding to any face) or as specific as a response to a particular face. For example, Perrett, Rolls, and Caan (1982) found temporal cortex neurons that responded selectively to visually presented faces, as did Young and Yamane (1992) and Leopold, Bondar, and Giese (2006). Fried, MacDonald, and Wilson (1997) showed similar sensitivities in the hippocampus and amygdala. Kreiman, Koch, and Fried (2000) also described temporal lobe neurons that responded selectively to categories such as "faces, natural scenes and houses, famous people and animals." Similar "invariant" responses to faces (among other stimuli) have been described by Quian Quiroga, Reddy, Kreiman, Koch, and Fried (2005) and Quian Quiroga (2012).

7. Although the idea of sensitivity to specific faces or objects dominates much of the research into single neuron coding, other selectively sensitive neurons encoding other cognitive processes have been recorded in various regions of the brain. Jenison, Rangel, Oya, Kawasaki, and Howard (2011), for example, report the observation of single amygdala neurons in awake humans that were linearly related to the "value" assigned to the options in a decision-making task. This was done by comparing psychophysical decision-making behavior with neurophysiological recordings from single neurons. Value, in this case, was used as a metric of the desirability of the alternative choices in a purchase decision. The authors claim that their findings were:

> . . . consistent with the hypothesis that single neuron activity in the human amygdala might play a role in either encoding or computing stimulus values at the time of choice.
>
> *(p. 336)*

8. Another popular single neuron theory of cognitive representation, but from the motor rather than the sensory system, is based on the concept of the mirror neuron. Acting on some early findings (di Pellogrino, Fadiga, Fogassi, Gallese, and Rizzolatti, 1992; Gallese, Fadiga, Fogassi, and Rizzolatti, 1996) it was suggested that single neurons in the F5 region of the prefrontal motor cortex respond not only when a monkey performs a specific action (such as reaching for an object) but also when it sees another monkey or a human perform that same action. The response of the mirror neurons was said to "embody" the perception of the motor action.

Subsequent research defined some special conditions and limitations for this selective response, but the important point in the present context is that a single cortical neuron was deemed to represent the significance of a complex behavioral response. Implicit in these observations is a theory of motor representation that is comparable to the sensory and perceptual ones previously discussed—a single neuron, when activated, could encode a complex motor experience. As with other single cell approaches to explanation, the mind–brain problem would essentially be solved should this interpretation be correct, since motor actions of virtually any complexity would be produced by activation of the appropriate neuron.

9. A number of "single neuron theories" seem to conflate single neuron and population activity. Sevush (2006), for example, has proposed a single neuron theory that ascribes "consciousness" to individual neurons. These neurons do not directly contribute to consciousness according to him but, acting in concert with other neurons with which they are interconnected through their respective dendritic trees, their joint activation is tantamount to the concomitant production of conscious psychological activity. These interconnections, according to Sevush, solve the binding problem (how we aggregate the isolated responses of the nervous system into a unified cognitive experience) by relegating it to the action of the dendritic interconnections. A close reading of Sevush's "single neuron theory," however, suggests that it is not very different from other network theories. He notes that "the production of conscious behavior results from population activity at the network level" (p. 705), which is also a principle of his theory.

Sevush also called attention to an even more microscopic problem—how does a neuron learn to respond to the selective stimulus sensitivity? His answer is that selective sensitivity is the outcome of "quantum entanglement" processes acting at the level of a single elementary particle. This is an idea also supported by Argonov (2012). Many of us would consider this idea unsupported by empirical evidence.

With this brief sampling and review of single neuron theories, we can now turn to a critique of the single neuron approach to the neural coding of cognitive processes.

2.4 A Critique of Single Neuron Theories

As the decades have passed by and many experiments have been carried out to study the mind–brain relationship, there is no longer any doubt that there exist neurons in the brain that are highly selective in the range and complexity of stimuli to which they will respond. This selectivity seems to be specific enough so that neurons can be found that respond to individual faces or objects or even to complex ideas and concepts such as freedom or altruism. Evidence of an extraordinarily precise selective sensitivity has been provided by a number of investigators as evidenced in the preceding section. For a complex of reasons, a theoretical approach has emerged that argues that the relationship between cognition and neural activity is instantiated in the activity of neurons that are highly selective to particular stimuli or whose responses are tantamount to experiencing highly specific concepts—that is, that are the "psychoneural equivalents" of mental experiences. Some of the factors that have engendered this kind of single neuron theoretical thinking include:

1. There is a well-developed technology with which the activity of individual neurons has been studied. Much has been learned about the physiology and anatomy of these neurons.
2. There is, as yet, little understanding of how cognition is encoded by the nervous system. There is currently no overarching neuroreductionist theory of cognition.
3. Thus, there is an explanatory vacuum attracting a variety of variously supportable and plausible theoretical options. This vacuum has been partially filled by cognitive neuroscience theories that dote upon extensive knowledge of single neuron physiology.
4. Single neuron theories are conceptually simple and plausible from a cognitive neuroscience point of view.
5. Because we want and need some explanation of how neuronal activity becomes cognitive activity, there are strong extra-scientific reasons to use whatever empirical evidence exists, no matter how sparse.
6. There is ample empirical evidence of a high degree of selectivity of the specific stimulus conditions that will activate a given neuron. This has to be distinguished, however, from any Gedanken experiment in which it might be shown that the activation of a particular neuron results in a particular cognitive activity.
7. There has been substantial success in understanding the neuronal codes used by the peripheral sensory and motor pathways to convey information to and from the brain. Specifically, the codes used by individual neurons to transmit the dimensions of stimuli are well known. Unfortunately, transmission codes are not the psychoneural equivalents of cognition, and success in the former field does not presage success in the latter.
8. Early work showed that there was a progressive convergence in the complexity of information represented by single neurons as one ascended the sensory transmission pathways. This convergence of geometrical properties was used as a heuristic for convergence of abstract ideas and concepts.
9. Some of these analogies were misinterpreted as necessary and definitive explanations of the neural representations of cognitive processes rather than just sufficient alternatives.

How these neurons arrive at their high degree of stimulus selectivity is presumed, by analogy to Hubel and Wiesel's geometrical properties in the periphery, to be the result of progressive convergence of less selective stimulus properties from one level of neural representation to more specialized and selective ones at higher levels. Because it is not likely that specific sensitivities such as a "rabbit," "Albert Einstein," a "yellow Volkswagen," or any other entity never previously encountered in the course of evolution, could be genetically programmed by evolutionary processes, it is usually assumed that some degree of adaptive or reinforced learning is also an important part of the process that determines the selective stimulus sensitivity of the neuron.

Two Interpretations of "Gnostic" Neurons

Two ideas, in particular, are regularly conflated in dealing with single neuron theory. We can characterize them as two postulates, one empirical and one speculative.

> Empirical Postulate A: Neurons can be observed in the higher reaches of the nervous system that respond selectively to complex and abstract stimuli. An example would be the selective activation of a neuron by the presentation of a particular face or even a political idea.

> Speculative Postulate B: The activation of a neuron tuned to selectively respond to a complex stimulus (such as an individual face) produces the psychological experience of that face.

There is no remaining doubt that individual neurons can be observed that respond very selectively to complex stimuli—faces, houses, tools, ideas, ideals, and so on. Thus, Empirical Postulate A is supported by an abundant literature justifying its being considered as an empirical fact rather than a speculation or hypothesis. Barring some subtle technical artifact mitigating the validity of these observations, Empirical Postulate A seems incontrovertible—neurons with highly specific sensitivities exist.

However, experimental support for the Speculative Postulate B—the assertion that correlated psychophysical responses will occur when particular neurons are activated—is a far more contentious matter. The experimental task of uniquely linking the activity of a particular neuron to a particular cognitive response is fraught with technical and empirical difficulties. Such an experiment would require locating and controlling the same neuron previously demonstrated to respond selectively to a particular concept, object, or face and then controlling it. This a far more challenging task than locating neurons that respond selectively to complex stimuli. It would then be necessary to demonstrate that the appropriate psychological response is causally associated with manipulation of the state of that particular neuron. Experiments directly speaking in support of Speculative Postulate B have never been carried out and probably never can be carried out. The relationship between the neurophysiological and the psychophysical responses would be supported only by functional analogy and not by some more robust kind of causal linkage.

Thus, we must ask—is the response of one of these high-level cognitive neurons just a passive indication of what had been a series of information processes that ultimately converged on that neuron? In this case, the correlation of the stimulus and the neuronal response would be cognitively irrelevant and tantamount to committing Yule's classical error of attributing causation to correlation. The specific and selective neuronal response, in such a case, would merely be an epiphenomenon of huge amounts of information that converged on a particular neuron.

The idea of a stimulus-correlated neuronal response without cognitive significance is a very different interpretation than one based on Speculative Postulate B—the idea that activation of such a neuron is tantamount to experiencing its meaning. In that context, a specific cognitive neuroscience (neuron to cognition) connection would have been made as opposed to just a neuroscience connection (stimulus to neuronal activity). To put these questions in other ways—is the response of a single neuron, which is tuned to a complex concept, the necessary code for the experience of that concept? That is, is it the psychoneural equivalent? If the answers to such questions are affirmative, we would have come a long way toward the solution of the mind–brain conundrum. Unfortunately, since Speculative Postulate B is beset by practical considerations, effective testing of single neuron theories of cognition seems unlikely in the near future.

Troubles in the Motor System—the Mirror Neuron

The mirror neuron concept is so fundamentally attractive as a psychological theme that it has been applied to what is a fair sampling of almost everything psychological. It reflects a widespread and persistent acceptance of what used to be called a motor theory of cognitive embodiment. Most famously, it has been used as a single cell explanation for "action understanding"—a somewhat ambiguous term for the ability to predict future actions and thus to understand "cognitive" activities. It is a theory providing a neural foundation for current and future motor activities by identifying neurons whose responses are associated with purpose and execution of motor activity.

However, the concept of a single "mirror" neuron as a necessary representation of a complex cognitive process has come under strong criticism in recent days. The difficulties are both empirical and conceptual. Efforts to find some correlate of mirror neurons using fMRI techniques have so far produced negative results (see the meta-review by Turella, Pierno, Tubaldi, and Castiello, 2009). The most compelling, persuasive, and far-reaching critique of the concept of single mirror neurons representing motor activity, however, has been offered by Hickok (2009). He lists "Eight Problems for the Mirror Neuron Theory of Action Understanding in Monkeys and Humans." This authoritative analysis should be read by everyone who is interested in single neuron theories but also to understand the logical complexities of mind–brain theories in general. Hickok's "problems" include:

- There is no empirical evidence in monkeys that mirror neurons support action understanding.

- Action understanding can be achieved via non-mirror neuron mechanisms.
- M1 [primary motor cortex] contains mirror neurons. [This means that the mirror neuron idea is "nothing more than the facilitation of the motor system via learned associations."] (p. 1237)
- The relation between macaque mirror neurons and the "mirror system" in humans is either non-parallel or underdetermined.
- Action understanding in humans dissociates from neurophysiological indices of the human "mirror system."
- Action understanding and action production dissociate.
- Damage to the inferior frontal gyrus is not correlated with action understanding deficits.
- Generalization of the mirror system to speech recognition fails on empirical grounds.

(Hickok, 2009, pp. 1231–1238)

This is not the place to expand upon the details of each of these arguments, but there are some general inferences that can be extracted from Hickok's analysis of the frailties of this particular single neuron theory of motor behavior. These include the following (after Hickok, 2009):

- The primary criticism Hickok raises of this approach is that it is not supported by empirical evidence. A considerable amount of subsequent research has failed to find the associations between perception of motor action and neural responses.
- The hypothesis of "action understanding" neurons is not unique. The idea of "mirror" neurons is not necessary and thus Ockham's razor should be invoked.
- Many of the so-called supportive experiments actively dissociate critical relationships between neural and behavioral parameters.
- When these neurophysiological findings failed to replicate, the crutch of mirror "systems" as opposed to single neurons was invented to patch up the failing theory.
- Despite these problems, the idea of neural representations of action understanding persists. (Action understanding is a new version of traditional motor theories of cognition in which experiences are tied to actual but subliminal motor responses.) This reflects the age-old problem faced by psychology of poorly defined attributes or faculties.

All in all, single mirror neurons as the psychoneural equivalents of cognitive processes appear to be another one of those popular fads that will persist despite strong arguments against both the basic concept and the empirical findings.

The Exclusivity Question

Another cogent counter-argument against single gnostic or concept coding neurons arises when one asks—if the activation of a single "pontifical" neuron is capable of creating an overall, unified cognitive response, what are all of the other neurons of the brain doing when this pontifical neuron is dominating our thoughts? In other words, how do we turn all of the other stimulus selective neurons off so that their responses are not also striving to produce equally compelling mental experiences? A more likely basic assumption, of course, is that vast numbers of neurons arranged in a network are most likely to be collectively involved in defining our thoughts. Obviously, there is a disconnect between the concept of the singular all-powerful "pontifical," "gnostic," or "concept" neuron, on the one hand, and the reality of the huge multitude of other neurons that are known to be active at any moment in defining our mental states, on the other. I refer to this as the exclusivity question. Unfortunately, the exclusivity question is also probably unanswerable and may be a "bad" question emerging as an unintended consequence of the original postulate of single cell coding. However, this is not the end of the story. There are many other criticisms that challenge the conceptual as well as the empirical status of single neuron theory. A few additional challenges include:

1. The conceptual leap from single cells to cognition is too long. It seems illogical that the vast and intricate neuronal networks should not play a key role in representing complex ideas.
2. Neural network theories are actually simpler than single neuron theories given the perplexities and unavoidable complications that follow single neuron theories.
3. Despite the huge number of brain neurons, it seems unlikely that every object of the external world would have its own neuronal representation.
4. Even if they did, the question of how selectivity for objects and concepts that did not exist emerges during evolution remains salient.
5. Many cognitive processes do not have simple topological relationships to the attributes of neural responses and, therefore, simple analogical isomorphism is not likely to be a satisfactory criterion in determining the psychoneural equivalent of a concept such as "democracy." Furthermore, there are some stimulus dimensions, such as stimulus chromaticity and responses such as perceptual color experience, that cannot be linked together directly by any kind of straightforward dimensional isomorphism. As a result we must be able to search for symbolic relationships and interactions—information processing mechanisms that are executed by the configuration of complex networks rather than by single neuron states.
6. Perceptual responses are dependent on strong interactions between both the dimensions of the stimuli and the meaning of the perceptual experiences; interactions of this kind are not accounted for in single neuron theories.

7. The sensitivities reflected in the neuronal responses are actually not that good a match to the concepts being encoded. As noted, what appears to be a face coding neuron may actually be one responding to primitive shapes (e.g., the t-shape implicit in the eye-nose-mouth configuration) that are inherent in face-like stimuli.
8. Artifacts are common—artifacts that are produced by partial properties of the stimulus.

Finally, like so many other examples of cognitive neuroscience research, replications of surprising results are rare. As experiments have been repeated, many of the putative relations between single neurons and complex cognitive processes, faces, or real objects have disappeared. In other words, the empirical basis for single neuron theories is neither abundant nor reliable.

2.5 Interim Comment

As provocative as these challenging questions and problems for the single neuron theory of cognition may be and despite the gradual evolution of the neuroreductionist zeitgeist away from pontifical and gnostic neurons toward neuronal networks, there is a residuum of interest and support for single neuron theories of cognition.

Single neuron theories, in general, are purported to represent an extreme "sparse" version on the continuum between localized encoding in the nervous system and distributed models of cognitive representation involving large numbers of neurons acting collectively. "Sparse encoding" is another way of implying that localized neural activity represents a concept with a relatively small number of highly specialized neurons. Distributed encoding theories propose, on the other hand, that very large numbers of unspecialized neurons working in concert and probably involving large portions of the brain are the essential psychoneural equivalents of cognition. If the latter, it would be extremely difficult to find a particular neuron that represents a specific idea because the salient neural network is composed of widely distributed neuronal activations, no one of which represents the entire concept.

Although the sparse single neuron theory is initially conceptually simpler and promises to provide a superficially easy opening to the solution of the mind–brain problem (one neuron–one concept), I have shown in this chapter that it is also beset with a number of conceptual and technical difficulties. Not the least of these is the problem posed by the rhetorical question—if a concept is encoded by the activation of a single neuron or single class of neuron, what are all of the other active neurons doing? Should the "one neuron–one cognition" hypothesis be correct, what is the meaning of all of the other neuronal responses simultaneously being executed when our thoughts are seemingly committed to only one track?

The "distributed" (i.e., neuronal network) theory (which is based on the idea that neurons in many parts of the brain are responsible for our cognitive experiences) does not have such a simple expression—from its perspective, the simple one (neuron) to one (concept) idea is incorrect at its most basic level. Be not misled, however; network theories confront us with the problem of how to empirically analyze the actions and interactions of the multitude of neurons that must be operating in concert to represent a particular concept. This poses a combinatorial problem of network organization that is very likely to be computationally intractable—but by involving all brain neurons in the process, it finesses the "exclusivity" problem.

Single neuron theories also generate a false need for some kind of binding process in which the responses of many neurons must be melded into the single unified experience. However, when one reevaluates this idea, it seems obvious that the distributed network hypothesis requires no such binding; all neurons are incorporated and required to collectively produce the psychological response. The state of the network—the configuration of all of the neurons involved in the network—is itself the psychoneural equivalent. No longer do we have one neuron equaling one concept but in its place we have the state of an extended multi-neuron network that offers few opportunities for a detailed analysis.

As a result, research attention has gradually shifted away from single neurons encoding both simple and complex properties to networks and macroneural brain locations as new devices (such as the microelectrode array and the fMRI, respectively) have offered alternative measures of nervous system activity and its relation to cognitive processes. Because network theories of various kinds represent the current consensus, in my opinion, most contemporary cognitive neuroscientists look upon "grandmother" neurons and their ilk as antique curiosities. One reason they have not received more attention is that many of the critical experiments are not executable in practice and in principle. Should one be interested in considering single neuron theory further, reviews by Sevush (2006) and Bowers (2009) present the opposing view. Our attention now shifts to the antithesis of the single neuron theory—the technology and theory that have developed to deal with multiple neurons interconnected in networks of varying degrees of complexity.

Notes

1. This is an excellent example of how the basic chemical properties of a molecule can lead to the production of organized cells and tissues. Any philosophy that assumes that there is a neural foundation for mind must assume that such natural processes also guide organic evolution.
2. Although the electrical activity of a neuron is a useful indicator of its activity, it is important to note that these electrical signals are really epiphenomenal. The real mechanisms of neuronal activity are chemical, specifically ionic and molecular, which

create electrical actions as the chemical activity takes place. It is far easier to measure electrical activity than it is to measure chemical activity and, thus, much of the discussion of neuronal action is presented in electrical terminology.

3. There is considerable controversy concerning the first use of glass, electrolyte-filled microelectrodes. Galambos (1996) suggests in a history of neurosciences in his biography that "glass micropipettes" may have been in use as early as 1910 (p. 187). Thus, their use was certainly being explored elsewhere around the time Gerard and his group published their methods.

3

MICRONEURONAL NETWORK THEORIES—TECHNIQUE AND METHODS

What is a microneuronal network theory of cognition? In its most extreme form, it would be a complete statement of the structure and activity of each and all of the relevant neurons involved in a cognitive process. It would show that particular neuronal networks are both necessary and sufficient to produce a thought or experience. Activating the neuronal network would presumably produce an experience and, then, if the neuronal network changed, that experience would change. In some ideal experiment, we would be able to identify the state of each neuron and demonstrate exactly the role it plays in creating the mental experience. The language would be a low-level one phrased in the scale of neurons and synapses.

Obviously, this ideal conceptualization of a microneuronal network is not achievable—any hope for such a detailed microneuronal "explanation" or "theory" of anything mental dies quickly on the altar of simple numerousness. Not even the most powerful conceivable computer would be able to process all of the possible combinations and permutations of an appropriately sized and interconnected neuronal network. Thus, in the past, we have turned to different kinds of prototheories such as the macroneural and single neuron theories just described in Chapters 1 and 2. Metaphors, analogies, and simple recapitulation of the observations take the place of complete, explanatory theory as we turn to what in hindsight are only simplistic and trivial explanations. We construct "theories" that are partial, explain nothing, or simply reiterate the basic principles of neural function. It is an unavoidable conclusion, therefore, that a complete explanation of how mental activity emerges from a tangle of interacting neurons is likely to remain the greatest mystery in modern science.

However, there are some principles on which most cognitive neuroscientists agree. The main one is that mind is probably the result of the operation of matter (and, most probably, microneuronal network organization and activity).

This is the basic ontological premise of modern cognitive neuroscience. That we can prove this assertion seems increasingly unlikely. However, this does not mean that we cannot learn anything about the mind or the brain by continuing to pursue some current paradigms. Psychologists have added enormously to our understanding of behavior, and neurophysiologists and neuroanatomists have explicated the nature of the components of the complex networks that make up the brain in a way that produces deep insights into what may be going on in the great networks of the brain.

Some of these insights come in the form of ever deeper metaphors for mind–brain relationships expressed in the language of other sciences. I now propose another metaphor—a much reduced version of what is probably going on in the brain but one that clarifies some of its basic organizational properties. I call this the "register" metaphor, a term borrowed from computer science.

First, what do I mean by a "register"? A register in a computer is a group of binary elements that function together to select one of a large set of alternative outputs. For example, consider Figure 3.1. This register consists of 12 simple elements, each of which can be either in the "on" or "off" state. Computer engineers call them bits. This means that there can be 2^{12} or 4096 different output states defined by the 12 input bits. In other words, a very small set of input components can determine which of a greatly expanded range of possible outcomes will occur. Yet each input component maintains its own integrity and is used in many different configurations of the 12 components. By itself each element tells us nothing about the state of the register, but as a member of the group it is very powerful in defining a large range of outputs. Each bit in this register is, thus, functionally connected to every other bit and participates in defining the output of the register, yet is still independent.

Assume that that there are not 12 but millions of neurons in the brain's "superregister." Assume further that each of these components of the register can have 1,000 or more states defined by its synaptic inputs. Somewhere in this

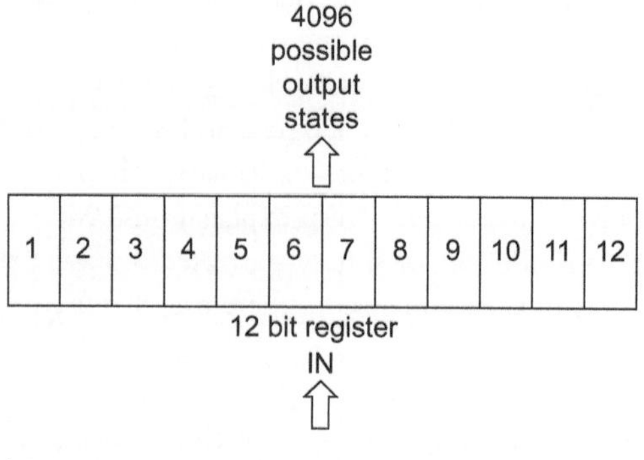

FIGURE 3.1 A 12-bit register can represent a much larger number (4096) of output states.

innumerable universe of possible brain states, there are emerging processes of which we are aware only in the form of experience—of human sentience or by observing the behavior of others. The fundamental postulate of the register idea is that even if the number of neurons is small, the number of states they can specify is vastly larger. Furthermore, given that the actual number of neurons is itself large, a much larger number of configurations and states can be encoded with the register than with any kind of single neuron. From one point of view, the superregister model of the brain is a surprisingly complete metaphor for what is happening in the brain. It incorporates many of the ideas that are widely accepted these days for individual neurons and for interactions among those neurons. It emphasizes a number of important aspects of mind–brain relations such as the role of the states of activity of multiple numbers of individual neurons. It does so without invoking any supernatural concepts or introducing any incompatibilities with any physical law.

It also instantiates some other ideas that seem relevant. For example, the role of any component may not be constant. Instead, what a "bit" means varies with the task being confronted by the register. Likewise, what a bit means is dependent on the role of all of the other bits in the register. Thus, there is a considerable increase in the complexity and number of concepts—a multiplier that extends the possible outcomes far beyond even what the 86 billion or so individual neuronal states might encode. A further source of uncertainty is that individual neurons may participate in several different networks at different times and in different tasks. Finally, the register system is robust; if a bit fails the system may still work—at least partially.

However, from an alternative point of view—practicability—register theory is, by its most fundamental nature, incapable of complete fulfilment as a theory of mind. The number of neurons involved in a realistic cognitive process raises the impossibility of dealing with all of the details of the neurons and synaptic connections and units of this vast network. As such, the most hopeful prognosis—specifying the exact configurations and states of the components of the brain's superregister that embody a cognitive process—is beyond any hope of realization. Thus, we are confronted with the unhappy conclusion that the main goal of cognitive brain theory—"breaking the neuronal code"—is not possible unless there is some kind of an intellectual and conceptual breakthrough not currently imaginable.

This conclusion may be interpreted to mean that we cannot go beyond what are essentially partial and incomplete metaphors of mind–brain transformations that exist today. Despite the fact that superregister theories must be correct in general principle, our analyses of them are currently constrained to simple toy models (i.e., reduced versions of the brain's actual neuronal network). In addition, they suffer from the same problem that all other theories of mind–brain do. They finesse or eschew the most basic problem of all—how does intangible mental activity arise from tangible neural activity? On this point, there is no argument—we do not yet know how brain becomes mind. This lack of progress, fortunately, has not obstructed thinking about neuronal networks. The starting point for a consideration of a microneuronal theory of how the brain works is Donald Hebb's famous conjecture, the next topic in our discussion.

3.1 Hebb's Conjecture

One of the most amazing, insightful, and creative examples of evidence-poor, but widely accepted theoretical speculations about the way in which neurons become mind is Donald Hebb's (1949) formulation of what we would now designate as a pioneering microneuronal network theory of cognitive processes. The essence of his conjecture was that cognitive states are encoded, implemented, or instantiated in the brain by the individual states of and interactions among neurons as mediated by variations in the strength of synaptic junctions. The synaptic variation is regulated by the use or disuse of those junctions in networks that he referred to as cell assemblies and phase sequences. In Hebb's words:

> Any frequently repeated, particular stimulation will lead to the slow development of a "cell-assembly," a diffuse structure comprising cells in the cortex and diencephalon (and also, perhaps in the basal ganglia of the cerebrum), capable of acting briefly as a closed system, delivering facilitation to other such systems and usually having a specific motor facilitation. A series of such events constitutes a "phase sequence"—the thought process.
>
> *(p. xix)*

This conjecture, based as it was on almost no empirical evidence other than what was then known about the nature of individual neurons, was a milestone in physiological psychology—the name by which cognitive neuroscience was known at that time. In presenting his schema, Hebb explicitly rejected other attempts to deal with psychological problems. In particular, he disavowed any interest in a neuroreductionist behaviorism or classical stimulus-response schema. What he offered, in their place, was the possibility of an explanation of cognition in terms of very specific neuronal concepts and measures. To understand how profound his contribution was, it is desirable to list the basic properties of his formulation:

1. All aspects of mind, cognition, consciousness, or whatever else you may wish to call mental activity, are the results of interneuronal processes and mechanisms. These processes and mechanisms are to be found at the microscopic level of neurons, not larger anatomical units.
2. Specific patterns of neuronal activity are the "psychoneural equivalents" of specific cognitive processes. Specification of the detailed state of participating neurons produces a particular cognitive process.

3. Hebb's formulation is one of action and interaction of a complex system of component parts—a network.
4. These component parts are, in Hebb's formulation, neurons (cells), a constituency that defines the conceptual level of his model as a microneuronal one.
5. These mechanisms are not unique in that ". . . alternative pathways each having the same function . . ." (p. 74) may exist. The system is therefore redundant—multiple neurons, different cell assemblies, and phase sequences can produce the same response.
6. The neuronal mechanisms of cognitive processes involved are distributed throughout the brain, the only exceptions being sensory and motor regions and the ". . . so-called speech area . . ." (p. 284).
7. The basic neuronal correlate of change in cognition is the establishment of changes in the efficacy of the interneuronal connections as a result of use. In Hebb's words:

> When an axon of Cell A is near enough to excite a Cell B and repeatedly or persistently takes part in firing it, some growth process or metabolic change takes place in one or both cells such that A's efficiency, as one of the cells firing B, is increased.
>
> *(p. 62)*

8. The "growth process" or "metabolic change" mechanism of changing interconnectivity between neurons is the increase in synaptic conductivity or efficacy, specifically in the growth of "synaptic knobs."
9. It was not known at the time of Hebb's conjecture exactly what was the functional equivalent of "change" in interneuronal efficiency. However, an increase in synaptic efficacy is now often referred to as Long-Term Potentiation (LTP)—the relatively persistent increase in synaptic conductivity that occurs following repetitive stimuli. (We nowadays have a much more complete understanding of how synaptic conductivity is modulated at the molecular level.)
10. There was no allusion by Hebb to specific localized areas, but brain-wide equipotentiality was not assumed.
11. Specific neurons do specific jobs in specific cell assemblies.
12. The system requires great numbers of neurons.
13. "There is no good evidence" for some of these ideas, nor do they provide "a means of testing" them (p. 66).

There is no question that Hebb's conjectures about cell assemblies were stimulated by the outburst of neurophysiological research in the first half of the 20th century, especially by such notable neurophysiologists as Lorente de No (1938) and Warren S. McCulloch (1944). It is somewhat more problematic whether World War II developments in information theory and electronic circuits had any direct effect on his conjecture. Notwithstanding the difficulty of tracing the origins of ideas, there is no question that his integration of a number of these ideas into a coherent conjecture was a monumental and historic event for neuroreductionist psychology. Hebb's conjecture, modestly modified, still represents today what most cognitive neuroscientists consider to be a working schema for the otherwise still unanswered mind–brain question. It is surprisingly modern and almost all of these statements would find broad agreement despite the fact that some of them (e.g., number 6) run against the current zeitgeist.

Thus, it should come as something of a shock for people who are not at the core of cognitive neuroscience thinking to realize how sparse is the empirical support for Hebb's conjecture—that cognitive processes are the result of a particular kind and arrangement of microneuronal activity as modulated by variations in synaptic conductivity. Although plenty of evidence shows that synaptic efficacy changes with use (LTP), only recently has some information appeared (Nabavi et al., 2014) that links synaptic efficiency changes to behavior such as learning. The history of the widely accepted, but scantily supported, Hebb conjecture has been described in detail by Takeuchi, Duszkiewicz, and Morris (2013). In that article they also make clear the paucity of direct support for the conjecture. In general, therefore, the gap between behavior and Hebbian neuronal networks remains obscure. However, this is what makes Hebb's "conjecture" just that and not an empirically robust "theory." As important as it was in the evolution of thinking about the relation between neurons and cognition, there is a real question about whether it is testable. We now examine this important question.

Is Hebb's Conjecture Testable?

Microneuronal interactions are the bases of the Hebb conjecture (Hebb, 1949)—a conceptualization that assumes that cognitive processes emerge from the idiosyncratic action of millions, if not billions, of individual neurons and the thousandfold greater number of synaptic connections; not in terms of their cumulative, pooled, aggregated, or summarized values, but instead to the degree they maintain their individual identity in the neuronal network. The main problem with the Hebb hypothesis, despite its logical force, conceptual robustness, and near universal acceptance, is that it has almost no empirical basis given the practical problems preventing simultaneous and independent manipulation of a cognitively significant number of neurons.

To establish the validity of the Hebb model would require an unobtainable amount of individual control over an uncountable number of neurons. Consider how a microneuronal experiment would have to be carried out. Ideally, the brute force technique of manipulating each neuron would require that we accomplish the following preliminary instrumentation steps:

1. Identifying each relevant neuron and all of their interconnections that are involved in a given cognitive process
2. Individually stimulating each of these identified neurons with a coherent pattern of energies tailored to each neuron's needs and roles in the relevant network encoding the cognitive process
3. Recording individual responses from every neuron to determine its individual and specific response in the relevant network

Obviously, carrying out these tasks would be a technical tour de force of the first magnitude given that our ability to develop microelectrode arrays suitable for stimulation or recording has so far been limited to a few hundred (e.g., Wark et al., 2013). Furthermore, the amount of data obtained by even a few hundred microelectrodes would tax available storage capabilities.

Should some extraordinary technical development allow us to carry out these demanding experiments, we would then have to determine how well the criteria for associating cognitive and neurophysiological measures proposed by Martin, Grimwood, and Morris (2000) are satisfied. Meeting their four criteria would authenticate that a microneuronal network response is the causal representation of a cognitive process; in other words, a theory of how mind emerges from brain.

1. Detectability: Stimulate the microneuronal network by driving it with a cognitive stimulus. Measure the pattern of neuronal network activity that is elicited by this "stimulus."
2. Mimicry: Activate the previously recorded pattern of neuronal activity and determine if it reproduces the cognitive process previously used as a stimulus.
3. Anterograde Alteration: If the pattern of network activity is blocked from forming, the cognitive process should also be blocked.
4. Retrograde Alteration: Changes in the neuronal network state should produce corresponding changes in the cognitive process.

(Paraphrased from Martin, Grimwood, and Morris, 2000)

Obviously, for a host of current procedural and instrumentation reasons, the technological tasks enumerated in the past several paragraphs cannot be carried out. It has not proven possible to record simultaneously and independently from anywhere near the number of identified neurons that are almost certainly involved in a cognitive process. Nor are we able to selectively stimulate an identified group of individual neurons to determine if we can mimic, block, or change a cognitive process.

To decode a cognitive state into a specific neuronal network, the implication of Hebb's first assumption, would require that Martin, Grimwood, and Morris' second mimicry criterion be satisfied. That is, controlled replication of the specific synaptic conductivity pattern (and the resulting specific settings of the individual neuronal state) should produce a specific cognitive state.

Because of these technical and procedural limitations, the brute force, microneuronal strategy of stimulating and recording from a realistically large network of individual neurons is probably not implementable and can only be conceived of as a hypothetical Gedanken experiment. Even if by some technological magic we were able to produce the necessary technical instruments, constraints on data accumulation and analysis would overwhelm any conceivable computer solution to the mind–brain for simple combinatoric reasons. Thus, there is a glaring inconsistency between the microneuronal level at which neurons really operate and the macroneural level of data obtained at the level of current brain-imaging technology.

However, the second part of Hebb's conjecture—that it is synaptic efficiency that controls the state of the network (i.e., producing particular cell assemblies and, thus, particular cognitive states)—has been tested, in general if not in individual neuronal detail. Indeed, this is what has been accomplished by Nabavi et al. (2014) in an extraordinarily important experiment during which this group activated and deactivated a brain circuit purported to control a behavioral response by means of genetic manipulations. Specifically they were able manipulate synaptic activity in the form of LTPs to establish a conditioned fear response and then reverse this "cognitive" state by producing a Long-Term Depression (LTD—a reversal of the potentiation) with different optogenetic stimuli. This was accomplished by using optogenetic stimuli to alternatively activate and deactivate synaptic activity, respectively. This procedure permitted Nabavi and his colleagues to enhance (LTP) or reduce (LTD) activity in the amygdala—a region of the brain associated with conditioned fear in the rat—by applying two different trains of patterned light as opposing stimuli.[1]

The extraordinary outcome of this experiment was that by using this technique, the rat could be controlled to exhibit a conditioned response analogous to fear under the LTP conditions, but behavioral passivity under the LTD conditions. Furthermore, and most interesting, was that these researchers were able to reversibly turn the fear behavior on and off by alternating between the LTP and LTD synaptic conditions. The proximal independent variable in this experiment, therefore, is the degree of conductivity of the collective responses of many synapses in a region (the lateral amygdala) of the rat brain thought to be associated with a well-known behavioral experiment: The dependent variable was a specific behavior. This experiment provided the best evidence yet obtained to support what only had been a hypothesis for many years—that the synaptic activity regulated the behavior, thus supporting the Hebb conjecture.

Although we still do not know the exact synaptic or neuronal states at the individual neuronal level for this model cognitive process, Nabavi and his colleagues have now shown that manipulation, albeit collective, of the degree of synaptic potentiation produces correlated (and possibly causally related) changes in an observable behavior. This is solid support for the parts of Hebb's conjecture that state that, in general, synaptic processes regulate neuronal states and thus behavior. Given all of the challenges in this complex field of research, this is a

remarkable achievement. This experiment meets the first, third, and fourth criteria enunciated by Martin, Grimswood, and Morris. The one criterion that it does not satisfy is mimicry, which requires information about specific patterns of individual synaptic conductivity—information that still remains elusive.

What Nabavi et al. accomplished, to sum up their conclusion, is that synaptic states can alter behavior by turning on and off a particular synaptic configuration, thus confirming one of the prime assertions of Hebb's conjecture. In their words:

> Thus, we have engineered inactivation and reactivation of a memory using LTD and LTP, supporting a causal link between these synaptic processes and memory.

Of course, this does not speak to the other part of Hebb's conjecture—that is, how the detailed pattern of the neuronal and synaptic states are decoded into specific cognitive states. That would require meeting the mimicry criterion, a test that is not applied in the Nabavi et al. experiment and may not be possible to satisfy.

Why should it be so difficult to provide an answer to the first part of Hebb's conjecture and to develop a microneuronal network theory of cognition? The answer to this question is that some of the very properties that make the Hebb conjecture plausible and logical also make it intractable to conventional laboratory experimentation. These properties include:

1. Microneuronal Level: The small size of the constituent components (neurons)
2. Numerousness: The large number of neurons involved in any cognitively significant neuronal network
3. Mechanical Obstructions: Practical limits on the way in which recording and stimulating electrodes can be brought to bear on neuronal tissues
4. Complexity: The rapid growth in the number of alternative arrangements of only a few neurons, much less the number of involved neurons in the simplest conceivable realistic network
5. Unpredictability: Responses are not deterministic but are stochastic; randomness would kill such an experimental approach even if the practical difficulties just mentioned could be overcome
6. Intractability: The dearth of mathematical and computer methods for solving even what seem to be relatively simple network problems, much less those corresponding to a "thought"
7. Underdetermination: Multiple explanations exist for any observed macro response
8. Redundancy: There may be no single answer to the question of psychoneural equivalency
9. Hidden Causal Influences: Indirect interneuronal connections may simulate direct connections

The only partial exceptions to these generalities are to be found in simple invertebrate "model" preparations such as used in the work on *Limulus* (Hartline, Wagner, and Ratliff, 1956), on *Aplysia* (Kandel and Tauc, 1964; Kandel, 1991) and *Drosophila* (Vogelstein et al., 2014), or in the many examples of well-structured peripheral sensory mechanisms such as the somatosensory cortex (Schwarz et al., 2014) or the retina (Litke et al., 2004; Margolis, Gartland, Singer, and Detwiler, 2014). It is in these locations or model preparations that the neuronal networks are simple enough for us to have detailed knowledge of their arrangement and how stimulus properties are specifically represented by neuronal states. Of this topic, I have much more to say in Chapter 4.

3.2 Microelectrode Arrays

A long-term motivating goal of neurophysiologists and cognitive neuroscientists is to observe the activities of neuronal networks. The conceptual basis for attempting to meet this goal is the so far unverified (but also the most plausible) conjecture that the cognitive information processing capabilities of the nervous system are embodied in the action and interaction of networks of microscopic neurons. One way to approach the goal of recording from many neurons is to alter our investigative strategy, which has hitherto emphasized single microelectrodes, and strive to develop Multiple Electrode Arrays (MEAs) to observe the activity of many neurons simultaneously. The hope is that by observing the activity of many neurons, we may be able to understand how neuronal networks operate and then to understand how the brain accomplishes its amazing functions. Whether the mechanisms and analyses of the situation will ultimately permit us to study simultaneously enough neurons to understand their relation to cognition is an empirical question to be resolved in the future.

Although the future is unpredictable, we must acknowledge that the current ideal of experimental manipulation of the vast numbers of neurons and their synaptic connections involved in cognitive processes is obviously still a wish rather than an accomplishment. Although we may learn an enormous amount about interactions of a few neurons or the specifics of synaptic growth, it seems unlikely given the enormous number of neurons, the redundancies, and the irregular and idiosyncratic nature of the brain's neuronal networks, that the necessary criteria of the kind proposed by Martin, Grimwood, and Morris could ever be satisfied.

This is not to deny that, at the present time, a considerable amount of progress has been made in developing MEAs that can record from hundreds, if not thousands of active neurons. This approach, however, poses its own challenges; some are simply computational and arise because of the many ultramicroscopic electrodes themselves that must be monitored at once. There is no assurance that even an MEA with thousands of electrodes would be sufficient to solve the problem of how neuronal networks account for cognition. Indeed, as we shortly see, just increasing numbers sometimes produces its own challenges. The need for the ability to examine the function of many neurons simultaneously is obvious given the polyneuronal basis of current neuronal network theory.

However, there are several caveats that have to be made before we embark on a discussion of the development of MEAs. First, no matter how many electrodes may be combined into an MEA, the number of electrodes (and, thus the number of neurons from which we may record) will probably always be smaller than the enormous number of neurons probably involved in any cognitive process.

Second, as we succeed in developing the technology to record from more and more neurons, we will be facing an increasing counterforce—the increasing combinatoric complexity and skyrocketing computational and storage requirements to analyze even modest-sized networks.

Third, despite some ingenious use of miniature manufacturing techniques originally developed for computer chip fabrication, MEAs, in general, consist of electrodes that are still too large to isolate the ultramicroscopic details of a cognitively significant neuronal network. This is a technological issue that may be overcome in the future, but, for the moment, represents a serious barrier to progress.

Fourth, MEAs, which are almost universally extracellular, consist of electrodes that connect to surrounding neurons in a way that is currently not predictable or regular, but at least quasi-random. Furthermore, even if the connections were deterministic rather than stochastic, it is by no means certain that the intricately interconnected neuronal network of many neurons would map directly from the MEA to the nervous system. Distant neurons may produce large responses that are incongruent with the geometrical array implied by the MEA. Alternatively, nearby ones may produce fortuitously small responses for what are usually obscure reasons where one would have expected large responses. Thus, the goal of inferring neural network organization from the observed responses may not be achievable—we may not be able to solve the clustering or sorting problem (to be introduced shortly) in which responses are associated with specific neurons or to achieve our goal of determining the connections between particular neurons. The problem of limited detailed knowledge of electrode-neuron connectivity is exacerbated by the dynamic, transitory nature of neuronal networks.

In general, we do not have adequate control over which neurons will be connected to which electrodes. Thus, associating the geometry of a particular neuronal network with cognitive processes remains a challenging task. Therefore, MEAs are likely to tantalize us by providing data that is conceptually relevant (cognition *is* a product of a microneuronal network) but that will never be able to fully satisfy criteria such as those proposed by Martin, Grimwood, and Morris (2000) for robust empirical proof. Make no mistake, we are certain to learn wonderful things about both the nervous systems and cognition, but, for the moment at least, what seems currently promising may just be another dead end in our quest to explain how brain matter becomes or produces intangible experience. For these reasons, it is unlikely that even the most advanced MEA will ever be able to provide a foundation for a detailed explanatory theory at this microlevel, however promising the idea may be. It may be necessary to conceptualize the problem in some completely different way that we cannot currently anticipate to break the neuronal network–cognitive code.

Although it is currently impossible to simultaneously record from enough of the enormous number of neurons that must be involved in even the simplest cognitively relevant network, efforts have been made to construct assemblages of electrodes so that the activity of a larger number of neurons can be recorded simultaneously than is possible with a single microelectrode. Currently available MEAs consist of anywhere from a few to several thousand electrodes. The guiding presumption is that by observing the sequence of activations or correlations among a modest number of neurons we may achieve some insight into the basic properties of more realistically sized neuronal networks. The following section details some of the technological progress that has been made in producing MEAs in the past decade or so.

Types of Microelectrode Arrays

Penetrating Microelectrode Arrays

Multiple electrode arrays come in two main types. The first is designed to penetrate into tissue or chunks of tissue; the second is designed so that more or less isolated neurons are laid on top of a printed array of electrodes. An example of the first type simply binds a number of individual wire electrodes together into a bundle (Figure 3.2). Because this type of MEA is designed to be thrust into neural tissue, it must consist of mechanically robust shafts. These electrodes are particularly appropriate for in vivo preparations in which the normal arrangement and metabolism of the neural tissue must be preserved. Electrode arrays of this type typically pick up a mix of signals (e.g., local potentials, spike action potentials, and electronic noise), depending mainly on the random geometrical relationships between the sampled neurons and the MEA. The individual electrodes making up the MEAs are designed so that certain positions (the tip and, in a few instances, other points along the shank) can be left uninsulated and thus make electrical contact with parts of neurons.

The simplest and earliest kind of penetrating electrode was a brush-like device in which individual wire electrodes were simply bundled together (Kruger and Bach, 1981; Nicolelis and Chapin, 1994). Preparation of the bundled wire electrodes is minimal. Other than stripping the insulation to expose an electrical contact, and possibly plating the tips of the several electrodes with a metal such as gold, silver, or carbon black for improved contact between the nervous tissue and the microelectrode, little more need be done to prepare the individual wires beyond what has already been described for individual microelectrodes.

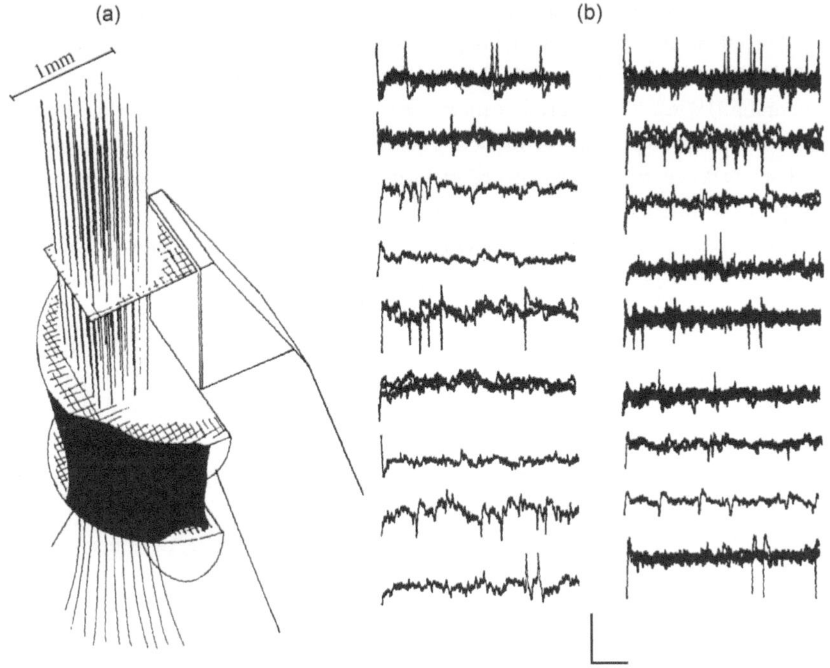

FIGURE 3.2 An early form of a multiple electrode array (MEA) formed by bundling a group of wire electrodes (a) and the typical irregular and mixed responses recorded with such a bundle (b).

Reproduced from Kruger and Bach (1981) with permission.

Figure 3.2A shows the arrangement of 30 bundled microelectrodes as constructed by Kruger and Bach. Figure 3.2B shows a sample of the responses recorded by them from 12 of their 18 electrodes. As was typical in all MEAs, not all of the bundled electrodes picked up a neuronal signal. It is important to note the irregular nature of the responses resulting from "noise" of various kinds obtained with these early MEAs. A major problem, both with regard to these early records and with later improved techniques, was that it was difficult to reliably determine which neuronal responses went with which neuron. The idiosyncratic heterogeneity of the responses picked up by simple "bundling" electrodes of this type is clearly evident. This problem, then, engendered a problem special to MEAs—the lack of a direct association of neural responses with their source neurons. We encounter this problem repeatedly in the following discussion.

The wire "bundle" electrode, however, does have one very important advantage—it is robust enough that it can be thrust into the brain of a living animal for "in vivo" recording. Thus, it can record from nervous tissue under relatively normal undamaged conditions. Its main disadvantage is that the individual wire electrodes are irregularly mapped onto the brain tissue. A compromise between a brush MEA with its unavoidable irregular electrode placement, on the one hand, and more mechanically robust and regular MEAs, on the other, was developed by a group at the University of Utah (Nordhausen, Maynard, and Normann, 1996). This array has many of the advantages of the bundle electrode but also overcomes many of its deficiencies.

The Utah electrode was constructed on a silicon substrate by a micromachining process similar to that used to produce computer circuits that etched away unwanted material. These electrodes had some important advantages in that they could pick up signals from a regular two-dimensional space and potentially also in three-dimensional space, if active electrode points could be set up on the shafts in addition to the one at its tip. Another advantage of the Utah array was that these electrodes were robust enough to remain inserted in brain tissue for periods that could be as long as 13 months (Rousche and Normann, 1998). An example of this kind of electrode is depicted in Figure 3.3.

Effort is currently aimed at developing MEAs in which many of the multiple electrodes in the array actually penetrate into the neurons, thus producing much larger intracellular electrical signals than was possible with extracellular MEAs. If this complex engineering task can be accomplished, then intracellular signals orders of magnitude stronger than extracellular ones can be recorded; this may open the door to MEA studies of synaptic potentials as well as the spike action potentials (SAPs) now being routinely examined. All of this depends, of course, on not being overwhelmed by the demands of the enormous amount of data typically generated by experiments using MEAs.

Improving the capability of penetrating MEAs has been the goal of other investigators. Kibler, Jamieson, and Durand (2012), for example, developed a

FIGURE 3.3 An MEA formed by a silicon micromachining technique consisting of 100 electrodes.

Reproduced from Nordhausen, Maynard, and Normann (1996) with permission.

method for producing high-aspect (in which the length of the electrodes considerably exceeds its width) MEAs so that the electrodes could be pushed deeply into slabs of mouse hippocampus. This large penetrating depth was required because the neurons of interest in their experiments were located deep in the brain. Kibler and his colleagues used arrays consisting of as many as 64 electrodes. Spira and Hai (2013) present a comprehensive discussion of these new developments in penetrating multielectrode technology and the problems they raise, as well as their advantages for the simultaneous recording of multiple intracellular responses.

Planar Microelectrode Arrays

The second type of MEA does not depend on penetrating more or less intact neural tissue. Instead, it is designed to record from neurons that have been isolated from their neighbors by dissection or chemical means. The idea when using this alternative is to print a planar set of electrodes and allow them to make surface contact with either dissociated or cultured neuronal preparations as shown in Figure 3.4. This figure is of special interest because it shows not only the 36 printed microelectrodes (the tips of the dark lines) but also some neurons and glia with which the electrodes are making contact.

Low-voltage, extracellular responses are the expected result of using this kind of printed electrode since the effective diameter of the individual electrodes in this kind of MEA is typically larger than the 1 μm diameter of glass or metal microelectrodes; they are therefore not suitable for use in in vivo intracellular recording but are primarily limited to in vitro extracellular recordings from neuronal tissue that is simply laid on top of them or excised tissue on which they can be placed. The responses recorded with such an electrode depend upon the uncontrolled physical relationships between the neurons and the electrodes.

Printed microelectrode arrays of this kind have been used to observe cellular activity since the 1970s (Thomas, Springer, Loeb, Berwald-Netter, and Okun, 1972; Gross, Rieske, Kreutzberg, and Meyer, 1977; Pine, 1980; Gross, Williams, and Lucas, 1982) and continue to be improved by current workers. The number of electrodes that can be fabricated continues to rise, with 64 or 100 being routine and 1,000 or more now being at hand. Indeed, Schwarz et al. (2014) reported a system based on cubical arrays of electrodes that could, in principle, record from as many as 1,792 electrodes, although random effects usually meant that far fewer were active at any time. The work reported by this last group was carried out on rhesus monkey primary somatosensory and motor cortices in which topological relations with the external world were preserved. As a result they were able to cluster neuronal SAP responses and to show that different areas of these peripheral regions of the brain correspond to different sensory and behavioral motions.

FIGURE 3.4 An MEA formed by a photoetching technique consisting of 36 "printed" electrodes. Neurons and other pieces of the mouse's spinal neurons can be seen in the background for scale.

Reproduced from Gross, Williams, and Lucas (1982) with permission.

However, there remain profound difficulties in using such an array. The simple act of mechanically connecting large numbers of multiplexed preamplifiers to such an array is a challenging task unto itself. As we see, most modern research is still limited to the analysis of relatively few neurons compared with the number of neurons that must be involved in even the simplest kind of cognitive process.

The main problem with such an MEA, as mentioned earlier, is not technological—indeed, advances in instrumentation, especially wireless transmission and microengineering procedures, are likely soon to overcome many or all of these handicaps. Instead, it is the vast amount of information generated by even a modest number of electrodes that may define the limits of their application.

What the future holds, of course, cannot be predicted and there has been continuing progress in increasing the number of available electrodes, if not the ability to process the large amount of data they could conceivably generate. In particular, Eversmann et al. (2003) produced an MEA consisting of 16,384 electrodes arranged in a regular array of 128 × 128 individual electrical contacts. Their MEAs were fabricated from complementary metal oxide semiconductor (CMOS) technology. Figure 3.5 shows a small portion of Eversmann's electrode array and some disassociated neurons that have been deposited on it.

To the best of my knowledge, MEAs of this size have never been used in a real experiment beyond limited demonstrations of the advanced technology

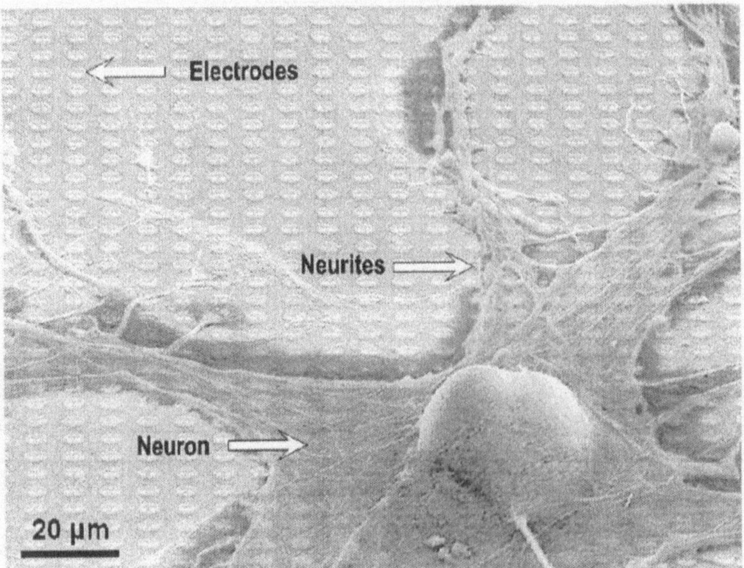

FIGURE 3.5 An MEA formed by a photoetching technique consisting of approximately 16,000 electrodes. A neuron can be seen lying on the MEA.

Reproduced from Eversmann et al. (2003) with permission.

developed by investigators such as Eversmann and colleagues. Nevertheless, MEAs of this magnitude suggest the possibility of a whole new approach to research on neuronal networks and contain the germ of applicability to cognitive neuroscience when combined with the advent of powerful supercomputers.

Steidl, Neveu, Bertrand, and Buisson (2006) have succinctly summarized some of the advantages that MEAs might have for neurobiological research:

- There is a need for micromanipulators to position electrodes in the tissue.
- Multiple electrodes can be used either for stimulation or recording.
- Printed MEAs are less sensitive to mechanical vibrations than are single glass microelectrodes.
- Printed MEAs can be used several times if adequately cleaned after each recording session.
- 3D-tip–shaped MEAs can pass through dead surface layers and record closer to living neurons within an in vitro slice.
- Parallel recordings at multiple electrode sites in a single slice of tissue provide the opportunity to observe region-specific effects, to increase individual data points and to improve statistical analysis.

(Paraphrased from Steidl et al., 2006)

As I noted earlier, it seems that the largest body of MEA-related research currently being carried out is aimed at engineering improved MEAs rather than neurophysiological or cognitive research per se. This emphasis includes research on topics that could generally be classified as housekeeping functions—for example, efforts to optimize the functions, simplify the construction, or avoid toxic effects of the materials used in MEAs. The following projects are examples of some of this kind of housekeeping research. The important message in this context is that although research on the design and construction of planar Multiple Electrode Arrays (pMEAs) continues, relatively little of this activity has yet been applied to the ultimate application (for cognitive neuroscientists) of these devices—understanding how neurons function in networks to produce cognitive activity. We can expect continued progress in improving the technology and in understanding interneuronal interactions. Whether these remarkable technological developments can be extended to solving the special problems of cognitive neuroscience is yet to be established. There has been a continued effort to develop improved pMEAs since the idea of using photolithographic techniques was suggested in the 1970s. (For a detailed discussion of current photolithography technology, see DeWeerth, Meacham, Giuly, Guo, and Hochman, 2008.)

Electrode arrays of those and earlier times were plagued by problems such as signal-to-noise ratios and tissue damage, not to mention that failure in achieving the intended raison d'etre for which these electrode arrays were being constructed: understanding how neural networks operate. The simple mechanical task of fitting all of the devices, amplifiers, conducting leads, and connections into a small region perplexed early investigators who attempted to construct MEAs. The eternal problem of impedance matching between the microelectrodes and the preamplifiers also continues to be a formidable obstacle.

Some of the specific challenges faced by engineers involved optimizing the spacing between electrodes, increasing the number of electrodes, increasing the sensitivity of the electrode-neuron interface, decreasing their toxicity, modifying electrode interface impedances, increasing the vitality of the neurons recorded by the MEA, and other such housekeeping functions. The work of Heer et al. (2004) exemplified the state of the art about a decade ago. Special emphasis was directed at two problems—improving the fabrication techniques and integrating the various electronic components.

Because of the complexity and expense of the standard photolithographic process used to construct pMEAs, a substantial amount of work is now being carried out to produce cheap and reusable MEAs by such workers as Charkhkar, Knaack, Gnade, Keefer, and Pancrazio (2012). Their technique deposited gold electrodes on a plastic (polyethylene napthalate) substrate. Their electrodes were insulated from one another by another plastic (parylene). These electrodes were robust and could maintain cultured neurons for up to 30 days and then could be sterilized and reused. As many as 64 gold-plated electrodes made up their version of an MEA.

Other researchers, such as Seker and his colleagues (2010), also concentrated their attention on the coating materials used at the active points on the electrodes in a pMEA. Their goal was to overcome poor impedance-matching conditions at the electrode-neuron interface. Gold was a fine contact material but efforts to use it were challenged by difficulty in making gold adhere to their electrodes. Seker et al. discovered that a form of porous gold was able to partially overcome this difficulty and not only adhered to the electrode much better than ordinary gold but also lowered the impedance of the electrode by a factor of 25. Similar reductions in electrode impedance and adherence were reported by Keefer, Botterman, Romero, Rossi, and Gross (2008) for thin layers of carbon nanotubes. This material is increasingly used today to increase electrode efficiency both in terms of lowered impedance and improved adherence to the surface of the microlithographically produced electrodes.

Because of the microscopic size of the apparatuses used in MEA experiments it is often difficult to provide a stable chemical environment in a perfusion chamber. Researchers such as Blake et al. (2010) have developed a micro-perfusion chamber that permits the necessary fluids to permeate both sides of a slab or slice of neural tissue, thus prolonging its viability.

Among the most fundamental issues faced by those who wish to use this technology is that there is no a priori one-to-one mapping from neurons to electrodes. Because of the irregular nature of the neuronal networks and the nature of the interneuronal environment, some electrodes would detect the activity of multiple neurons and some would detect nothing; a few would fortuitously record the activity of a single, identifiable neuron, but this is probably the exception rather than the rule. As we see later, one of the ultimate goals of MEA research is to develop analytical techniques to separate mixed neuronal responses and associate them with individual source neurons.

3.3 Applications of Microelectrode Arrays

"Housekeeping" research projects such as those just discussed are necessary to develop the MEA technology, a technology that makes many promises as the next best empirical approach to ultimately understanding how neural networks function. However, at present the state of the art in MEA technology has not yet developed to the point that we are able to parse the centrally mediated, cognitively related neural activities. Despite the fact that technological development is moving apace, it is unlikely that our ability to unravel such neural networks will soon be available simply because the relevant neuronal networks are without doubt going to be much larger and complex than we have hitherto been able to handle analytically.

Nevertheless, the next question we must ask is, beyond housekeeping and technology-based research discussed in the previous sections, what kind of neuroscience application-oriented research has been carried out so far with these remarkable tools? Although, as I noted earlier, the preponderance of the research being carried out these days is technology-directed, the field is not barren and a considerable amount of what we may appropriately call "pioneering" projects has been carried out in various fields in which the MEAs have been applied to specific neurophysiological problems. Accepting the constraint that few of these applications are even remotely cognitively related, most current applications deal with the simple geometry of peripheral sensory coding mechanisms or overall field effects in which information concerning the details of network organization is lost.

An outstanding and rare example of a cognitive neuroscience application of MEAs is the work on a sensory problem carried out by a collaborative group from several California universities (Litke et al., 2004; Shlens et al., 2009) who studied excised monkey retinae in vitro. The retinae were laid on top of a planar MEA consisting of 512 electrodes. To carry their experiments, it was also necessary for them to fabricate integrated circuits consisting of all the amplifiers, analog-to-digital converters, and multiplexers necessary to record from such a large number of electrodes. Using this system, the receptive fields were initially described by Litke and his colleagues for 364 identifiable neurons. The neurons were classified according to their on-off characteristics.

This research (Shlens et al., 2009) used the California integrated hardware system to study patterns of synchronized firing in an effort to explore the organization of the retinal networks. In particular, they observed that the functional response of retinal ganglion neurons depended largely on their immediate neighbors, and the retinal ganglion cells seemed to be arranged in a relatively simple anatomical pattern in which connections were limited to nearby neighbors.

Shlens et al. wisely concluded that their observation of the relatively simple pattern of organization in the retina would probably not be generalizable to more central brain structures. Citing the work of Yoshimura, Dantzker, and Callaway (2005) on the nonrandom network properties of the multilayer organization of cortical columns, they suggested that their ability to work out the detailed network interactions in a "simple" structure such as the retina may not be possible in more central structures.

MEAs have also been useful in determining the toxicity of various chemicals on neuronal activity by examining the overall diminishment of neuronal activity (Steidl et al., 2006). In their experiment, excised slabs of hippocampal tissue were perfused with antagonists that specifically inhibited synaptic transmitter activity. MEAs have advantages over single electrodes in such an application because they can measure system-wide effects in multicellular preparations that may not be obvious in single cell responses. Furthermore, the specific interconnections of the neurons from which they recorded were unimportant in light of their goal to determine a system-wide effect. A comprehensive and up-to-date discussion of the use of MEAs in neurotoxicity evaluations can be found in Johnstone et al. (2010).

The search for properties of central neuronal networks has been limited so far because of their complexity. Currently, in most cases, only the overall or average properties of a network can be determined or the data made available only in some other cumulative form. Although it is possible with current MEA instrumentation to observe the activity of a number of neurons simultaneously, it is far more difficult to identify the sequential dependencies and, thus, interconnections of their activities, and, therefore, the specifics of their network organization.

One very germane example of this kind of research that highlights both the advantages and disadvantages of MEAs is the determination of the bursting activity of cultured neuronal tissue. It is in this study that the capabilities of the MEA technique are currently most clearly demonstrated. It is also in this context that the great weakness of applying the MEA technique system is also made clear.

Bursting activity of neurons has been observed for many years (e.g., see Murphy, Blatter, Wier, and Baraban, 1992). It is particularly noticeable in in vitro clusters of cultured neurons. However, it can also be observed in intact slabs of neural tissue. Bursting seems to be a ubiquitous property of isolated neural tissue and may be due to an absence of inhibitory stimuli from other parts of the brain or, quite to the contrary, as a result of exogenous stimulation (e.g., see Wagenaar, Madhavan, Pine, and Potter, 2005).

The problem with any measurement of the activity of multiple neurons with MEAs is that it is difficult to determine the sequential order (or hierarchy of firing) in which the multiply interconnected neurons respond (a problem highlighted by Hilgetag, O'Neill, and Young, 1996). This is especially true in in vitro preparations in which virtually all of the neurons in a specimen may ultimately respond simultaneously and spontaneously. This near universal activity negates one of the most promising hopes of those who use MEAs—the expectation that they will allow us to examine the significant interactions of a network of neurons. As a result, although there have been some associations of bursting activity with behavioral learning, no obvious relationships have been observed that have plausibly related bursting activity to controlled stimulation. In fact, in one of the most notable (for reporting a negative result) articles on this problem using the components of an experiment we have just been discussing (MEAs, cultured neurons, and bursting activity) Wagenaar, Pine, and Potter (2006) reported no difference in the development of bursting activity when stimulated and when spontaneous. What we are left with in such situations is a dearth of information describing the detailed properties of neuronal connections. Others (e.g., see Belykh, de Lange, and Hasler, 2005) argue that synchrony of bursting neurons depends only on the general topology of the network rather than the specific interneuronal interactions. Specifically, they suggest that the synchrony of bursting brain tissue depends solely on the number of inputs to a neuron and not the specific details of the network. This interpretation suggests that "bursting" may be an abnormal endogenous condition that occurs when some threshold number of neurons is active. The obvious analogy, of course, is to seizure activity in intact brains.

The inescapable conclusion one must draw from the extreme instability exhibited by the bursting phenomenon is that not only is the brain complex, but also it is not stable over time. Brain networks seem to reorganize themselves for mysterious reasons in a way that may preclude any notion that we may be able to decode a consistent neuronal network, much less a stable experimental environment. Psychologists well understand the variability of our thoughts. Any experiment, therefore, that seeks to correlate cognitive and neurophysiological findings may be shooting at a moving target in a way that makes the goal of understanding the network of neurons that makes up cognition impossible.

A useful and practical recent application of MEAs has been their use as stimulators in sensory prosthetic devices that serve as supplements to deficient vision and hearing. Among the most successful applications of this technology has been in the auditory system in the form of chronic cochlear implants. Among the first to experiment with this idea in a surgical context were Djourno and Eyries (1957). In their procedure, electrodes of the kind shown in Figure 3.6 were inserted through the oval window directly into the snail shell–shaped cochlear space following along its curving course. At various points, an electrode shorn of its insulation permits electrical stimulation of the remnants of the cochlear nerve and any surviving hair cells.

Cochlear implants have been remarkably successful, as anyone who has spoken to a person with such a device will quickly appreciate. Evaluated behaviorally, the success of such a system is really remarkable and conversation can take place almost normally. Clinical evaluations by several laboratories generally show 85% sentence comprehension (see the review by Zeng, Rebscher, Harrison, Sun, and

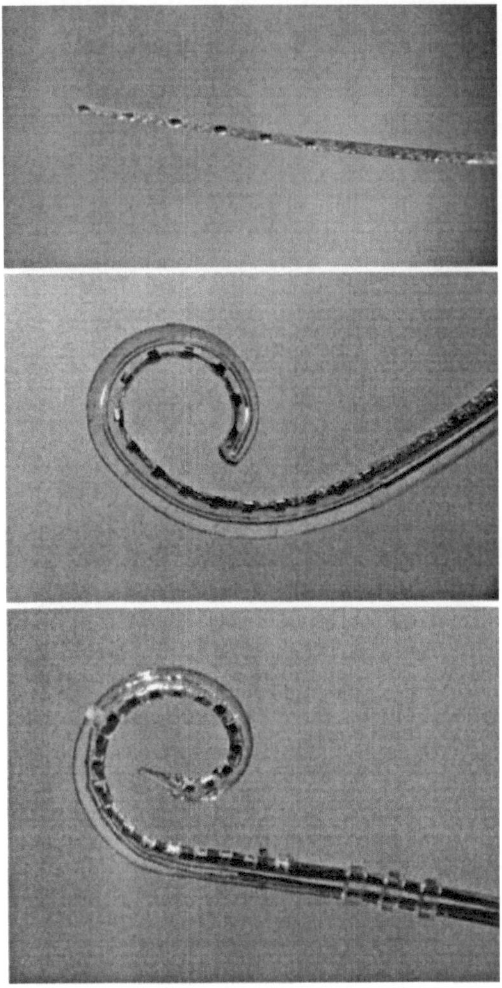

FIGURE 3.6 A prosthetic MEA consisting of approximately 20 electrodes used for electrical stimulation of the cochlea.

Reproduced from Zeng et al. (2008) with permission.

Feng, 2008), a notable improvement over patients who lost hearing due to hair cell and early acoustic nerve damage.

The high level of success with cochlear implants is largely due to the one-dimensional and orderly spatial manner in which the various acoustic frequencies are linearly encoded along the cochlea—a discovery for which Georg von Bekesy (1899–1972) received the Nobel Prize in 1961.

MEAs such as those shown in Figure 3.7 are also being pursued experimentally as a prosthesis for vision (for an up-to-date review see Luo and da Cruz,

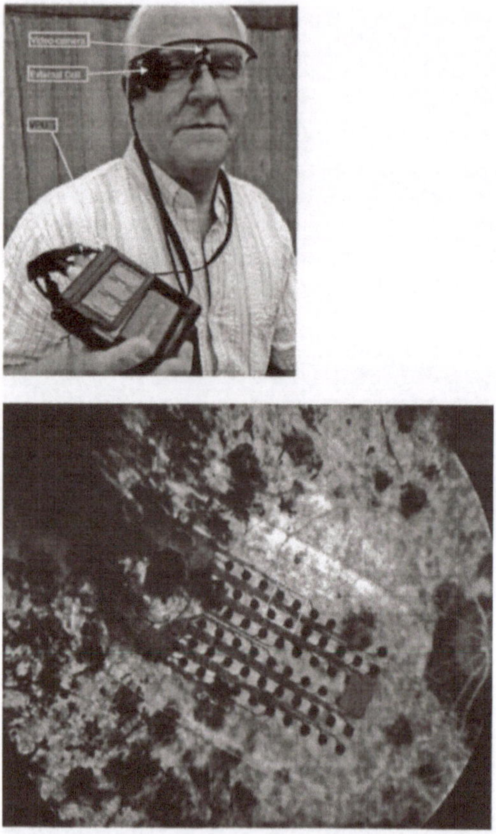

FIGURE 3.7 A prosthetic MEA consisting of 60 electrodes used for electrical stimulation of the retina.

Reproduced from Luo and da Cruz (2014) with permission.

2014). However, the ambitious goal of applying an array of microelectrodes to meaningfully stimulate many retinal neurons and reproduce functional vision of a quality comparable to the results achieved in hearing remains a yet to be perceived task. Perceptually, the few patients who have had this experimental treatment report "seeing" nothing more than simple configurations of spots of light. The main problem in producing artificial vision is the complex anatomy and two- and three-dimensional information coding used by the neural network of the peripheral visual system. The design of the visual prosthesis is hampered by the very small size of the retinal neurons and thus the mismatch between the resolving power of the best microelectrode and that of the neurons in the retina (Zrenner, 2002). Figure 3.7 is a current intraretinal MEA as used by Luo and da Cruz (2014).

3.4 Analysis Techniques

The bare fact that we can record the responses of a larger number of neurons with an MEA than with a single electrode does not necessarily mean that useful information about the organization of the great neuronal networks of the brain can be extracted from those recordings. Similarly, the ability to establish the simple interactions between pairs of neurons does not guarantee that the specific polyneuronal codes for cognitive processes, whatever they may be, can also be determined. Currently, there is a qualitative mismatch between what our experimental techniques can accomplish in mapping a relatively few components of neuronal networks and the number of neurons in the networks that must account for consciousness, behavior selection, learning and adaptation, and all of the other cognitive activities that psychologists have probed and measured over the years. How far we can go in pursuing this approach—analyzing (i.e., mapping) the organization of brain networks—to solving the mind–brain problem is unknown at present. The development of increasingly powerful computer techniques leaves the door open to some extraordinary new developments in the future. However, there are harbingers in the current literature that the task is going to be far more difficult, if not intractable, than it may seem at the present.

This section deals with some of the progress that has been made in analyzing neuronal networks using MEAs. The word "analysis" conveys different meanings to different investigators and, for this reason, it seems appropriate to consider what an analysis means in the present context. In its most pristine form, a complete analysis of a neuronal network (or, for that matter, any network) would be a map identifying not only the momentary state of each individual neuron involved in the network but also the determination of how each neuron is connected to any other. Network connectivity, in this case, implies that there is a causal (i.e., functional) relationship between specific neurons or groups of neurons; in other words, that a given neuron's state is influenced by many other neurons. This influence may be either positive, exciting the neuron's activity, or negative, inhibiting the neuronal activity. It may be observed functionally or anatomically, but in the final analysis anatomical connectivity must trump any inferred functional connectivity. Given that each brain neuron may have hundreds of synaptic junctions converging on it, serious problems arise concerning how one would represent, much less construct, such a functional map and how one might empirically confirm either its anatomy or its function.

The suggestion is that the goal of mapping the neuronal networks underlying Hebb's conjecture in particular and cognitively relevant neural networks generally is probably going to be an overwhelmingly formidable task. We can know in principle what an idealized version of such a network would be like—the metaphorical superregister introduced earlier in this chapter. Connectivity can be depicted for small networks by a simple map characterized by a collection of nodes (the neurons) and by lines (the connections) indicating the interneuronal

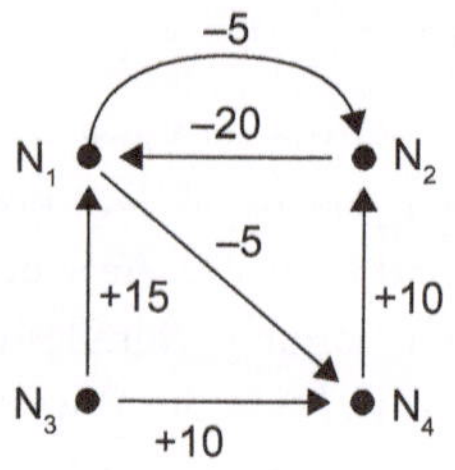

FIGURE 3.8 A simplified "toy" network consisting of four neurons, $N_{(1\text{-}4)}$. Arrows indicate connections between neurons and the numerals indicate the strength of these connections. Although relatively simple, this network is probably not amenable to prediction and solution.

connections. The valence and magnitude of the connections must also be represented by appropriate values if the network is to be fully characterized. Neurons not directly interacting would simply not be connected by lines, although this stratagem might obscure indirect routing of signals that would produce correlated activity among even very remote neurons in the network.

It is also generally assumed that such a map would have to indicate in some way that the constituent neurons would be dynamic, that is, capable of being in different states at different times (otherwise the network would not be functional, just locked in a constant state). The momentary state and the sequence of states are additional properties that have to be specified if a network theory of cognition is to be complete. Figure 3.8 is an example of the neuronal network for a highly simplified "toy network."

In this figure, signed numerical values indicate the potency of the interconnections between neurons (negative values being inhibitory and positive vales being excitatory) and the arrows indicate the direction of interconnections. At each moment in time, we assume that the ensemble of states of all of the neurons (N_n) in this hypothetical "toy" network would determine the network's function; at other times and other states of the component neurons, different functions would be encoded by different configurations of this same network.

Extrapolating this basic idea of a "toy" map to the brain would constitute a neuroreductionist theory in which the momentary states and interconnections of the network would presumably be associated with a particular mental process. Despite some limited progress with simple invertebrate nervous systems (where real networks contain a small number of neurons), extrapolation of the network concept to realistic vertebrate neural networks has not proven to be possible currently. The best we can do is speculate about general properties (e.g., neurons change their state as a result of changing synaptic efficacy).

It is also important to remember that the behavior of even as simple a toy network as the four-neuron example shown in Figure 3.8 may itself be unpredictable. Indeed, it is well known in mathematical circles that even a three-neuron network may create a situation that is the formal equivalent of the three-body problem in physics[2]—general solutions to which have long been known to be intractable under the conditions of feedback and interaction that are characteristic of all kinds of networks—biological and physical.

If even approximate, partial versions of cognitively relevant network maps are to remain the immediate goal, then it is necessary to answer a number of initial questions concerning the data obtained with the MEA technique by means of appropriate analytical procedures. The first question is—which neurons are members of the network being examined? Of necessity, this task is complicated by a number of issues. Trying to carry out such a task by repetitive probing with a single microelectrode would be tedious and ultimately unsuccessful because of the uncertainty that successive penetrations would make contact with the same neuron. Such an approach also ignores the loss of vitality of the neurons over extended periods of time. The great potential advantage of the MEA technique with its ability to simultaneously record from many neurons is that it promises to answer this first question directly, if not completely. However, just how far one can go with even 1,000 electrodes in identifying a set of interacting neurons remains problematic.

Next, consider that the extracellular potentials picked up with an MEA are not simply associated on a one-to-one basis with a particular electrode. Not only will the activity of individual neurons be detected by more than one electrode, but also one electrode will pick up the activity of several neurons. This raises the second basic analytical question—which of the recorded responses go with which of the identified neurons? This task is known variously as "response sorting" or "clustering"—the assignment of responses that may have been picked up by multiple electrodes to their source neurons. This may require the parsing and then assignment of responses from among the noisy mixture of responses recorded by each electrode. (See Figure 3.2B for a realistic example of responses picked up from an array of electrodes.)

The third analytical question that has to be answered is what are the specific functional interconnections and neuron states of the putative network that correspond in what is still a mysterious way to some cognitive or mental activity? This search for the neuronal network configuration or code is the "holy grail" of cognitive neuroscience. The complete specification of the relationships between mind and brain at this level would constitute fulfillment of the quest for a solution to the mind–brain problem. Unfortunately, for practical and conceptual reasons we now consider, this "grail" may remain as remote and unobtainable as the mythical one that Arthur's knights failed to find.

Spike Action Potential (SAP) Clustering

Consider the basic task of finding which of several extracellular responses picked up by an MEA can be segregated, sorted, or clustered in such a way that they can be associated with individual neurons. In some idealized sense, it seems that it should be possible to perform an initial segregation of neuronal response by concentrating on a single property of the neuron's response; for example, the shape of the spike action potential (SAP), its amplitude, or the interval between SAPs. In the latter case—SAP interval patterns—the suggestion is that we might be able to use periodic intervals as a clue to which responses go with which neurons. The basic postulate is that responses that are separated by equal intervals are the product of the same neuron. Unfortunately, if interval periodicity is the only clue, mathematicians (Cox and Smith, 1953) have determined that clustering of neuronal responses is extremely difficult to do because information about interval periodicity is lost when the intervals from a number of neurons are pooled. Mixing neuronal interval responses is known as superposition and the Cox and Smith superposition principle asserts that if intervals are pooled, the resulting pattern of intervals in the pooled output meets all of the criteria of *random* sequences! Thus, periodicity alone cannot be used to determine which responses go with which neurons, simply because measures of this dimension of the response (interval) are lost when SAP intervals are mixed. Cox and Smith (1953) proved that even if the responses of a particular neuron are perfectly periodic (i.e., their inter-SAP–intervals were all equal to some constant value), if one pooled the interval information from a number of neurons:

> . . . except in degenerative cases, the result is indistinguishable from that for a random series, no matter what the form of the individual outputs. In particular if the number of sources and the individual recurrence times are large the output will be random over intervals containing many events.
> *(p. 6)*

In a subsequent article that provided biological relevance to what otherwise might have seemed to be but a bit of mathematical esoterica, Cox and Smith (1954) showed that this restriction holds specifically for motor nerve ending response data collected by Fatt and Katz (1952). Their conclusion was that "there is no evidence that the [pooled] series is not completely random" (p. 99). What this means is that it is impossible to use residual interval periodicity alone to assign SAPs to individual neuronal sources—the intervals promptly converge to a random sequence for relatively small numbers of intervals.

There remain other questions about whether other single properties (such as amplitude or response shape) can be used to accomplish the SAP sorting task. However, an implication of the Cox and Smith work is that it is probably not possible to sort neuronal responses on the basis of any other single parameter except in the most reduced "degenerate" cases. For example, consider another simple SAP sorting procedure—one based on amplitude. Given the most reduced case in which a collection of SAPs have a spread of very different amplitudes and the signal-to-noise ratios are not themselves overwhelming, a simple measure of the amplitudes of recorded extracellular SAPs could conceivably be used to segregate a few responses as belonging to the outputs of a small number of individual neurons. Such a criterion might well work for a few neurons; however, should the signal-to-noise ratio be such that the amplitude of the noise is close to the amplitude of the signals, any such simple criterion is doomed to failure. Neuronal variability would quickly overwhelm any ability on the part of the analysis technique to assign detected SAPs to specific source neurons on the basis of the single parameter of amplitude.

Because it is unlikely that simple criteria such as SAP interval pattern or amplitude will work for more than a few neurons, other more advanced processing techniques are necessary to solve the sorting problem. Several more sophisticated analysis techniques have been suggested that involve measuring the shape of the SAPs—a technique that also depends on relatively noise-free conditions. In this technique, the shape of each of the recorded SAPs is digitized by an analog-to-digital converter. The resulting tables of numbers representing the shapes of the recorded SAPs are then compared to determine if they are similar enough to be assigned to the same source neuron.

Shape recognition of SAPs and their assignment to source neurons can, in principle, be carried out by a matching procedure in which the shape of a SAP is compared with the shape of a prototypical "template" response shape. The template may be a sparse one (a "window discriminator") in which only a few key elements from a template are used or a more complete one (a dense version—a "vector" discriminator) consisting of all of the digital values. Characterizing the SAP straightforward convolutional techniques may then be computed between the digital samples representing the SAP and those of the template; correlational values above some threshold are then considered to designate SAPs with common neuronal origin. This template-matching technique was in use as far back as the 1960s by such workers as Gerstein and Clark (1964) and continues to serve as the basis of much of the modern work in SAP sorting.

The SAP clustering task becomes increasingly difficult when multiple electrodes are used to study multiple neuronal responses. However, several advantages are obtained as the number of electrodes involved increases. For example, simultaneity of response can serve as a cue that the response recorded at one electrode is the same as one recorded at another electrode. In this case, any idiosyncrasies resulting from interneuronal distances become irrelevant. It is also possible, in principle, to begin to study the spatial arrangement of neurons when using multiple electrodes since response size is known to be strongly dependent on the distance between an electrode and a neuron. Unfortunately, all such techniques eventually fail at relatively modest numbers of neurons.

Other analytical procedures to assign or sort a mélange of SAPs to individual neurons have also used familiar statistical methods (e.g., Principle Component Analysis, Adamos, Kosmidis, and Neophilidis, 2008), new statistical techniques (e.g., Continuous Basis Pursuit, Ekanadham, Tranchina, and Simoncelli, 2014), or novel techniques such as wavelet transforms (e.g., Quiroga, Nadasdy, and

Ben-Shaul, 2004)—a method comparable to Fourier analysis but using different basis functions. Others who have tackled the problem with new techniques include Pillow, Shlens, Chichilnisky, and Simoncelli (2013). Their system, which they have designated as "binary pursuit," depends on a model of SAP shape that they claim improves upon sorting produced by the less effective earlier methods. A still timely review of the many published SAP clustering techniques has been presented by Lewicki (1998). This article, along with more up-to-date discussions of the topic (e.g., Brown, Kass, and Mitra, 2004; Wild, Prekopcsak, Sieger, Novak, and Jech, 2012) provide useful tutorials on the efforts being made in the development of techniques for sorting and clustering SAPs.

As formidable as these SAP sorting techniques appear to be at first glance, it is clear that there are many difficulties in applying them. Some of these difficulties are intrinsic to the problem (e.g., the loss of information that occurs as a result of pooling interval data) whereas others are matters of practical details (e.g., electrode movement induced variations in response properties occurring over time). Matters of mechanical and electronic technique are always subject to improvement and further developments are likely to overcome many of the technical instrumentation problems in the near future. However, some problems intrinsic to both the analysis methods and biological variability are probably unavoidable and not likely to be solved soon. It is a good idea to distinguish what the fundamental limits are and avoid wasting time and resources attempting to make the intractable tractable.

Lewicki (1998) summarized some of the difficulties that inhibit solutions of the SAP sorting problem. I paraphrase and update his list in the following comments:

1. Lack of independence of the neuronal responses. If neurons are interconnected in a way that their responses are not independent but highly correlated with each other (such that neuronal response X always leads to neuronal response Y), then the sorting problem becomes extremely difficult, if not impossible.
2. Nonlinear mixing of the recorded responses may distort various attributes of the SAP signal, including shape and the respective amplitudes of responses occurring at different distances from the recording electrode.
3. Lewicki especially emphasized the fact that any technique for SAP sorting that used the shape of the extracellular response is going to be severely degraded if the shapes are not constant.
4. As we have seen, neurons in vitro tend to produce spontaneous bursting activity—a response pattern that is likely to distort the shape of the SAP by exhaustion of its metabolic and ionic resources, thus producing various degrees of neuronal refractoriness.
5. Similarly, we can add that variability in virtually any other property that an experimenter might use to sort SAPs onto neurons will tend to make the

analysis quickly fail if the property is neither linear nor stable. Pillow et al. (2013) have been especially critical in noting the deleterious effect of overlapping SAPs on spike clustering techniques.

6. Simple mechanical and crowding problems of the ancillary signal recording and transmission equipment are a continuing problem. Neurons are microscopic entities; throughout the history of microelectrode use, both single and when enhanced by MEAs, investigators have had to fight the battle of making their instruments small enough so that they do not interfere with the target neuron mechanically. Furthermore, just getting the necessary parts of the recording system into the container holding the neuronal tissue remains a continuing engineering challenge.
7. Similarly, any nonrigid electrode configuration that is free to move about can lead to what Lewicki referred to as "electrode drift"—gradual changes in the spatial relationships of neurons and electrodes. This will inevitably lead to changes in the shape of the recorded responses and inhibit SAP sorting and clustering. Stabilization of electrodes has been greatly helped by the application of photolithographic techniques from computer fabrication engineering. Indeed, in most in vitro experiments, printed electrodes are perfectly stabilized in the containment well in which the (typically dissociated) neural tissue is placed. Unfortunately, because of tissue flexibility, stabilizing the relative position of the neurons and the electrodes either in vivo or in vitro experiments is still an unsolved, but serious, problem. Thus, the spatial arrangement of the interneuronal network is likely to fluctuate. Although some progress has been made in "caging" single neurons near single electrodes (e.g., Erickson, Tooker, Tai, and Pine, 2008), the problem remains a real one for those who wish to study neuronal populations over prolonged periods of time.
8. Lewicki also pointed out that with any communication system, noise is always recorded superimposed upon the signal information of main interest. Depending on the signal-to-noise ratio, critical information may be fully lost if this ratio is too small. Even worse, if the noise is not stable, changing properties from moment to moment would make the analysis process especially challenging. Indeed, unstable noise may completely obliterate any possibility of sorting SAPs using any of the measures mentioned here.

(Paraphrased from Lewicki, 1998)

The Futile Search for Connectivity Maps

Spike action potential sorting or clustering (assigning a SAP to a particular neuron) is not the only important task when one is attempting to study neuronal networks. Another major analytical goal is determining the connectivity or topological organization of a neural network. In other words, we must also ask—what are functional connections that permit one neuron's activities to drive or cause another's activity? Specifically, it is to determine how activity in Neuron X stimulates or causes activity in Neuron Y and ultimately to represent the organization of a network of many neurons that may account for a cognitive process. This is a far more demanding task than sorting SAPs to particular neurons simply because it requires a robust solution to the sorting problem before the connectivity issue can be attacked and solved. However, we must be repeatedly forewarned that beyond a very few neurons (a number that may be as small as three) determining the organization of a neuronal network or characterizing its neuronal connections may not only be difficult but impossible.

The problem is exacerbated by a vague, indeed, sloppy use of the word *connectivity*. It is not entirely clear that different authors mean the same thing when they use this ambiguous term. In an article that deserves much more attention than it has so far received, Horwitz (2003) pointed out that brain "connectivity" has a variety of different meanings. He distinguished between three main meanings—anatomical, functional, and effective connectivity. Anatomical connectivity is conceptually clear enough (and, when available, it is the "truth" standard against which the other two must be judged). Anatomical connectivity is what Sporns, Tononi, and Kotter (2005) referred to as the "connectome"—"a comprehensive structural description of the network of elements and connections forming the human brain" (p. 245).[3]

Horwitz defined functional connectivity as "the temporal coherence among the activity of different neurons . . . measured by cross-correlating their spike trains" (p. 467). The cross correlation measure was supposed to indicate the strength of an observed relationship between neurons.

Horwitz went on to assign a different meaning of connectivity in the third context—effectiveness. Effective connectedness is "the simplest neuron-like circuit that would produce the same temporal relationship as observed experimentally" (p. 467). From one point of view, this is the goal of all of the analytical methods that are described in this section—the actual map of the effective interconnections. He also made an exceedingly important statement at this point in his discussion. He argued that assaying the effective connectivity is severely limited because the neuronal network responses are underdetermined by the kinds of measurements we are able to make. That is, there are many possible neuronal networks that can produce the same functional neuronal activity pattern, and available experimental data can neither validate plausible nor reject improbable ones.

A similar vagueness permeates the nature of the word "complexity." Like its partner "connectivity," this word has different meanings to different authors and lacks a precise definition in the cognitive neurosciences. This makes it difficult to compare networks (although precise numerical values at the level of complexity with which microneuronal brain researchers must deal probably do not make any practical difference). Nevertheless, like the huge numbers associated with factorials or Bell's numbers, it would be worthwhile if such a numerical value for complexity existed. Unfortunately, mathematicians also disagree on a definition of complexity. It is often approximated by neuroscientists as a simple estimate of the number of neurons and synaptic connections in neural tissue, but this misses the whole point.

Network complexity, therefore, is viewed as a property of systems with numerous parts and inter-part connections. Measures such as the factorial can provide estimates of "complexity," but, in general, the utility of these numbers collapses as the number of components and interactions increases and numbers such as the factorial soar to hypercosmological levels. For computer scientists, the complexity of a problem has to be computable in order to lead to criteria for the solvability or convergence of certain algorithms. For cognitive neuroscientists operating at the microneuronal network level, almost everything beyond three neurons is, for all practical purposes, "infinitely" complex.

The reason for this uncertainty forewarns us about one of the most fundamental limitations of any technique that seeks to provide an effective map of a neuronal network. It is something that is likely to raise the hackles of those who seek to represent the structure of a neuronal network but it is essential that those who seek to use MEAs to disentangle neuronal networks understand its connotation, if not its denotation. Horwitz (2003) went on to say:

> A key experimental issue that shaped these definitions originated in the near-impossibility of knowing the exact anatomical relationship between a neuronal pair whose functional interactivity was being accessed (at least in mammalian experimental preparations).
>
> *(p. 467)*

Should this be true for a pair of neurons, how much less likely is achieving the goal of mapping the "functional interactivity" of a suitably sized collection of cognitively significant neurons? Thus, the quest for a unique effective map may always be a search for a chimaera; the best we will be able to do will be to develop a functional description—effective maps being denied to us.

Despite these musings, work continues to determine the organization of realistic neuronal networks—mostly with invertebrate model preparations. It is worthwhile reiterating that although some progress has been made in response clustering of simple preparations, the grand promise of producing a solution to the general network problem (mapping or decoding the specific neuronal network for a particular cognitive process) is still unfulfilled and nothing on the horizon is promising.

An interesting place to start a discussion of this kind is with a work that can only be described as a heroic example of a class project. This work is presented here because of its pedagogical simplicity and conceptual clarity. It is unlikely that this approach could prove to be an efficient procedure for mapping real neuronal networks, but nowhere are the methodological and logical assumptions behind neuronal network construction made clearer and the inevitable failure of the "holy grail quest" highlighted.

Two students at MIT, Jonathan Karr and Daniel Herman, authored[4] a laboratory semester project in which indications of microneuronal network connectivity were determined by the following method. Their experiment[5] used data recorded from an in vitro culture of hippocampal neurons using an 8 × 8 MEA. Each of the 64 electrodes was used in turn as a stimulator and the responses induced in all of the other 63 electrodes measured. Correlations were then individually computed between the stimulus applied to the single stimulator and the responses measured at all of the other responding electrodes—a selection that might be fewer than 63 because not all of the electrodes picked up a neuronal response. This produced a maximum of 4,032 correlations (64 × 64 minus the 64 autocorrelations of the stimulator with itself). The amplitudes of the correlations were interpreted to be measures of the degree of interconnectivity among the electrodes; high correlations, driven by high amplitudes and short latencies, being tantamount to intimate causal interconnections and spatial propinquity between the stimulating and the responding neurons.

Karr and Herman's brief report is exceptional not only because of its successes but also because of the way it highlighted some of the limitations of the whole idea of using MEAs to determine the connectivity of a network of neurons. On the positive side, their technique embodied the ability to provide initial estimates of a number of the connectivity parameters of the hippocampal neuronal network they were studying. Based on the assumption that the probability of an inferred connection would be higher for a correlated pair of electrodes than for an unstimulated pair and such other factors as the delay between the stimulus and the response, preliminary estimates could, in principle, be made of factors such as the geometry of connection (e.g., how far apart the stimulated and the responding neurons were either spatially or in terms of intervening neurons) and the interconnection strength (i.e., the magnitude of the correlation between the single stimulating and the other 63 responding electrodes).

Despite the unusual accomplishments of these students, their work also demonstrated the extreme difficulty of the task of determining the connectivity of a neural network with MEAs and the most powerful analytical techniques available. For example, there is a cryptic assumption in experiments of this kind that an ideal, pseudoregular neural network underlies these correlations. Neuroanatomically, however, the irregular geometry of a real neuronal network does not even come close to that of the regular 8 × 8 array of the microelectrodes used to collect these data. Collectively, the real neuronal networks are much more irregular. Whereas a realistically complex neuronal network may have dozens, if not hundreds, of synaptic connections between pairs of neurons and many thousands on each neuron, the analytically derived network implicitly assumes a much simpler kind of network based on a much simpler concept of connectivity than is actually present.

To exacerbate this situation even further, as I noted earlier, each extracellular SAP may be detected by more than one electrode and that one electrode may

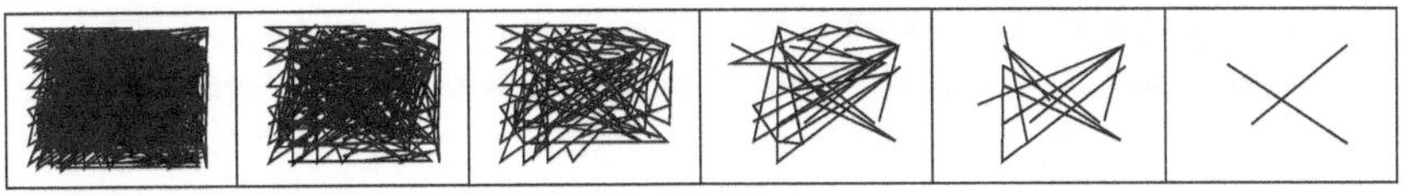

FIGURE 3.9 The effect of arbitrary thresholds on the connectivity pattern of a neural network.

From Karr and Herman (2004) with permission.

record activity from many neurons. Therefore, even under the best conditions, analyses with even the most limited goals are constrained to deal with small numbers of neurons and unverifiable ambiguities.

Another issue is the arbitrariness of the interconnectivity networks generated by methods like those reported by Karr and Herman. The main measure used in their analysis—correlations—as a measure of interconnectivity can be instantiated in a large number of different maps depending on arbitrary decisions. If a low threshold for a correlation is used as a criterion for interconnectivity, the resulting pattern would suggest a large number of putative links. On the other hand, if a high threshold was used, only a few links between neurons would be generated, as shown in Figure 3.9. The essential problem is the arbitrariness of the threshold for what would be acceptable as a connection.

Nevertheless, the importance of Karr and Herman's laboratory exercise went far beyond the preliminary nature of their accomplishment. Others (e.g., Maccione et al., 2012) also subsequently reported the use of correlational techniques similar to those used by these two students. Whereas Karr and Herman stimulated the electrodes to define correlated responses, Maccione and his colleagues allowed spontaneous activity to define causal functional relationships and thus to imply interneuronal connections. In addition, Maccione et al. also used filtering techniques to reduce the number of spurious connections—combinations of distance and latency being used to eliminate at least some of the implausible connections.

In recent years a number of other more sophisticated schemes have been developed that purport to determine the connectivity pattern of a neuronal network by means of MEAs and novel analytical techniques. One of the most comprehensive is the work of Stevenson et al. (2009). This group applied Bayesian methods to determine the statistical properties of neuronal networks in both toy and in vivo preparations of monkey brain neurons. Their goal was to characterize the functional connectivity of networks in the form of a matrix in which causal relations between pairs of neurons were either present or not. Their data were presented in the form of a matrix that showed which neurons were interacting with which others. Scores were binary values of interactions between all possible pairs of neurons in a manner comparable to the correlational scores reported by Karr and Herman. However, neither group of researchers was able to produce a connectivity map of the general type desired nor could they distinguish which interconnections were monosynaptic and which were the result of indirect pathways.

To accomplish what they did, Stevenson and his colleagues had to make assumptions about the nature of the neuronal networks they were studying. These simplifying assumptions ("priors" in Bayesian terms) made it possible to avoid a combinatorial explosion, but also limited how detailed the inferred networks could be. Their priors included, first, the assumption that neuronal interconnections are sparse, a generalization that almost everyone agrees is not anatomically true. Second, they also assumed that neuronal interactions are temporally smooth, that is, that the interactions occur without abrupt discontinuities in time. This assumption is also probably incorrect.

In summary, we now list some of the difficulties and challenges in reported attempts to determine the connectivity of a neuronal network:

1. At the base of many of the other problems faced by those who attempt to map out a realistic neuronal network is the fundamental underdetermination of the data obtained with an MEA. That is, it cannot be definitively determined by any analytical method yet proposed which of several (many?) equally plausible network configurations is responsible for a functional response. Neuron X may be connected to Neuron Y by a single direct synapse or by an indirect pathway that may involve a number of neurons. Any pair of neurons may be interconnected by a multitude of alternative network connections. Simply demonstrating a correlation does not adequately define a specific anatomical connection. Indeed, there is a strong possibility that it cannot, in principle, do so.
2. There is a discrepancy between neuronal networks determined electrophysiologically and those produced by analytical methods. It is often necessary to assume that there is a sparse pattern of connectivity (often approaching one-to-one connections) to avoid generating intractable combinatorial problems. In anatomical fact, however, all of our microscopic evidence suggests that each neuron is complexly and multiply interconnected with many other neurons, by as many as 1,000 synapses.
3. Unlike the neat grid-like arrangements of MEAs, neurons are irregular and idiosyncratic in both their shape and connectivity.
4. In many cases, real neuronal networks are influenced by neuronal activity from outside the network under study. Thus, both centrifugal and centripetal sensory or motor activity may be influencing what was originally thought to be a network that could be studied in isolation.
5. Almost all of the current methods are applied in experiments that, of necessity, report the use of relatively small numbers of neurons in their analyses. The availability of the MEA suggests, erroneously I believe, that this is no longer an issue since the activity of large numbers of neurons can be recorded simultaneously. However, a careful scrutiny of the literature suggests that the interconnectivity pattern of only a few neurons has actually been determined. In short, 1,000 electrodes in an MEA does not mean that the interactions

among 1,000 neurons can be determined. In practice, the neuronal networks reported in all research I have encountered so far are sparse and few. In general, the MEA technique has largely been applied to a few cultured, but disassociated, neurons in vitro; in far fewer experiments have any attempts been made to map neuronal networks. The basic reason for this constraint is the influence of biological combinatorial complexity, a topic to be expanded upon in the next section. For all practical purposes we are still constrained to work at the level of a very few neurons, usually in the nervous systems of invertebrate model preparations. This approach is discussed in Chapter 4.

3.5 Combinatorics of Complexity

Given the idiosyncratic nature of its interconnections and the large number of its components, an argument can be made that the brain is the most complex bounded entity in the universe. Regardless of which level of analysis one is considering, the combinatorial complexity of the brain stresses any putative microneuronal explanation of its function far beyond what we may predict are the limits of any modern or prospective science. The basic conclusion we must infer from experiments such as those just discussed is that it takes very few interacting objects or events to pose an intractable combinatoric problem. However, the nature, extent, and influence of neural network complexity are not a part of the intuition of many cognitive neuroscientists. The goal of this section is to fill that gap by introducing complexity and the limiting role it plays in cognitive neuroscience.

Early on, Stockmeyer and Chandra (1979), among many others, showed how relatively simple network problems (e.g., minimizing a traveling salesman's itinerary) could require hyper-astronomical processing times. For example, estimates suggest that if the salesman had as few as 21 stops, an exhaustive solution to the problem would require 77,000 years of computer calculation time.[6] Meyer (1974) proposes a general theorem that forcibly makes this point:

> THEOREM. If we choose sentences [binary strings] of length 616 . . . and code these sentences into 6 × 16 = 3696 binary digits, than any logical network with 3696 inputs which decides truth of these sentences contains at least 10^{123} operations. . . . We remind the reader that the radius of a proton is approximately 10^{-13} cm, and the radius of the known universe is approximately 10^{28} cm. Thus, for sentences of length 616, a network whose atomic operations were performed by transistors the size of a proton connected by infinitely thin wires would densely fill the entire universe. *(p. 481)*

Similar calculations can be made to show that there are many other problems that cannot be solved in the lifetime of the universe. Therefore, neither time nor space is adequate to deal with the complexity of even relatively (superficially) simple problems in the lifetime of the universe.

Karp (1986), a prominent complexity theorist, also reported how many other superficially simple combinatorial problems, many of which were analogs of those faced in cognitive neuroscience, were actually known to be intractable in their demands on time and space. Karp's (1986) work on the intractability of many other superficially simple combinatorial problems is especially relevant in this context. Specifically, he pointed out that many, if not most, real-world problems are said to be NP-complete, that is, they cannot be solved in real amounts of time. Although the work of Karp and others is well known in mathematics and computer theory, cognitive neuroscientists seem to be assuming that some of their greatest challenges are in principle solvable when it has already been proven that they are intractable.

Despite these reminders, there has been an implicit acceptance of the idea that the mind–brain problem will ultimately be resolvable at the microneuronal network level. (See the discussion in Chapter 5.) In this section, I consider some basic arithmetical background suggesting that any attempt to develop a connectionist model of anything approaching a cognitively significant neuronal network is not a feasible experimental goal.

By far the most profound challenge facing those who seek to determine the functional topology of real neuronal networks is what can be designated as "combinatorial or computational complexity." The general term "complexity"—a phrase thrown about by cognitive neurophysiologists with total abandon and freedom from any precise definition of the term—is heard in virtually every conversation about the brain. The term, regardless of how deeply we understand its numerical implications, certainly has profoundly affected our research by both explicitly and implicitly limiting our research efforts to toy networks with only a few neurons or to macroneural networks.

The constraint that inhibits progress in understanding network theories of all kinds is how large the number of alternative organizations of even a very small number of interconnected components can be. The basic arithmetical terms that are germane to the topic of complexity are *combinations*, *factorials*, and *permutations*. These terms are relevant to neuronal networks because the number of alternative, yet equally plausible, organizations of a very few components or nodes can very quickly rise to huge numbers.

Factorials (n!) are particularly relevant to neural networks because they correspond to the number of different ways that n objects (e.g., neurons) can be arranged (e.g., interconnected) into different groupings. The sum of all of the possible maps that are necessary to fully characterize a network quickly becomes enormous. The factorial n! of a number n is defined as the multiplicative product of all of the positive integers that are less than or equal to n. Even for small numbers, surprisingly large factorials are generated. For example, although 3! is equal to 6, 10! is equal to 3,628,800. Larger factorials (for example, 25! is equal to 1.551121004 × 1025) very quickly approximate and then exceed the number of basic particles in the universe. This numerical explosion provides a good idea of what happens even when one studies relatively simple networks.

Combination (C) is a more specific term that is defined in terms of the number of ways that a particular subset of objects can be arranged. The formula for combinations of n things taken r at a time is:

$$C = \frac{n!}{(n-r)!(r!)} \quad \text{Eqn. 3.1}$$

C can also get very large very quickly as n grows. Consider the classical example of playing ordinary 5-card poker for which n = 52 and r = 5. Applying Eqn. 3.1 specifies that there are 2,598,960 such combinations if one ignores the order in which the items are selected. Considering how much larger n and r can be in the brain, the number of coded neuronal combinations is also enormous.

If the order of the ways that a set of things are combined matters, then the word "combination" has a more restricted meaning—it becomes what is called a permutation. Because the particular way in which a set of neurons is functionally arranged would presumably lead to different cognitive processes, it is probably more correct to speak about changes in brain states as permutations rather than combinations.

Furthermore, this is not the worst case. There are other mathematical series that grow even faster than factorials or combinations—e.g., superfactorials—that although not a priori irrelevant to this discussion help to give us an intuition about brain complexity. A superfactorial of n is the product of the first n factorials. All such "super" functions make the same point: exhaustively studying networks or seeking to identify the unique arrangement that encodes a cognitive process is probably not currently a feasible or plausible approach to generally solving the riddle of how the brain makes the mind. Although we may be able to make some statements about the general properties of such networks, it is obviously impractical to attempt exhaustive experiments to understand their function. This idea of factorials as measures of possible organizations exceeds by far the already serious problems generated by the simple numerousness of the neurons in the brain.

The link of these esoteric mathematical concepts to neuronal networks is clear. Even if we were able to represent the brain as a system of billions of uniform and simply interconnected "billiard balls" it would underestimate the true complexity of the brain. Neurons are not simple entities as are billiard balls, but are themselves complex entities with many idiosyncratic properties and a variety of interconnections of varying weight.

Koch (2012) recently put this problem into perspective when he calculated that the time required to "exhaustively describe a [neural] system" (and this is what we assume we have to do to explain how the brain generates the mind) would grow faster than exponentially. As an example, Koch calculated that the 2 million neurons of the visual cortex of a mouse would require a computer system running for 10 million years even if the computer's powers were simultaneously growing at the modern rate suggested by Moore's Law ("computer component density and thus computational speed doubles roughly every 2 years"). Koch did go on to note that there were some simplifying conditions (comparable to those used to speed up a standard Fourier transform to a "fast" FFT version) that could speed up the process by a few orders of magnitude. One way to do this is by grouping neurons into clusters. However, whenever one regularizes a procedure in this manner, there is the ever-present possibility that the cognitively relevant microneuronal organizational properties one is seeking may be tossed out.

Koch (2012) also reminded us of another measure of complexity—Bell's number—defined as the number of different partitions into which a system of n objects can be divided. For example Bell's number for n = 10 is 115,975. Playing out its calculations serves as another reminder of the complex ways in which the components of a network can interact or be divided. A Bell's number of 115,975 compares to factorial 10 which is equal to 3,628,800. Thus, factorials grow faster than Bell's numbers (and both scale faster than exponentials) but the effect is the same—huge eventual demands for analytical speed if we are ever to exhaustively study the brain as an example of a neuronal network.

Whatever the best measure, they all pale into insignificance compared with the actual number of neurons in the human brain, and much of current thinking is that almost all of them participate in encoding cognition. The only hope, according to Koch, of ameliorating this constraint of "colossal" numbers of calculations is to find some way to reduce the number of subunits so that the number of possible partitions can be reduced. However, the irregular nature of most brain tissue (notwithstanding the discovery of such prototypes as repetitive columnar organization) makes it likely that we will join such alternative mathematical treatments as "graph theory" in attending not to the specific problems of cognitively realistic networks but rather to the general properties such as reachability, directedness, and degree of toy networks. As interesting as the work on graph theory of Harary (1969) or Christofides (1975) may be, little attention is paid by graph theorists to predicting the function of complex graphs or the differential effects of different partitions. Their attention is drawn to general properties of very simple graphs that may extrapolate in some ill-defined way to larger networks. This decision by graph theorists to concentrate on simplified abstract models is well chosen since, as noted earlier, even the three-body (or, in this case, the three-neuron problem) is probably not generally solvable.

"Complexity" has other ramifications beyond simply the raw numerousness of the nodes and connections on which I have concentrated so far in this section. Another aspect of complexity may be summed up in the term "noise." Noise, in the context of neuronal networks, refers to all of the sources of uncontrolled variability that tend to distort or destabilize response measures. This adds a level of temporal variability to the spatial kind of studies described in earlier sections of this chapter. It is difficult enough to deal with the topological (i.e., spatial) complexity of stable neuronal networks; how much more difficult it would be to deal with functional networks that are constantly changing their states and yet sometimes emitting the same behavior.

Obviously, for a host of procedural and instrumentation reasons, the experiments necessary to establish the cognitive role of a specific neuronal network are probably impossible to carry out. Experiments of this general kind, studying particular cognitively significant networks, make impossible demands on our technology, our mathematics, and our computer capability, not to mention the conceptual foundations underlying this approach. It is likely that we will never be able either to manipulate neuronal network states or measure the subtle differences between the aggregate neuronal states representing different cognitive states. Yet, this is the kind of experiment that would have to be carried out to "decode" or "map" brain representations of cognitive states. Proposed experiments at this level are at best Gedanken experiments that can be discussed ad infinitum but never consummated.

Others, such as Lichtman and Denk (2011), have also argued that any proposed attempt to study the brain at the microscopic level of size and, thus, at the enormous numbers of neurons, is likely to be beyond both our instrumentation technology and the necessary analytical capability of current mathematics. Their reasons include (1) the immense diversity of cell types in the brain; (2) the diversity of cellular responses in both space and time; (3) the scale difference between microscopic neurons and their extent over vast macroscopic regions of the brain (the brain must be studied "over sizes that span six orders of magnitude" [p. 620]); and (4) the overwhelming flood of data at the microneuronal level. Lichtman and Denk concluded that:

> There is also no other organ system for which the complexity of the structure is so great that earnest arguments can be made against delving into structural analysis because such an effort might well provide an unprecedentedly gigantic, yet totally incomprehensible, mass of data.
>
> *(p. 618)*

Although Lichtman and Denk express optimism that the task of mapping neuronal circuits may be tractable sometime in the future, as of this date I have not been able to find any report of a cognitive (i.e., something beyond a peripheral motor or sensory mechanism) linkage to vertebrate microneuronal networks that meets the criteria expressed by Martin, Grimwood, and Morris for robust empirical proof at the microneuronal level.

3.6 Interim Conclusions

The major conclusion of this chapter is that there exists additional profoundly compelling, if not overwhelming, evidence that suggests that the search for the specific microneuronal networks that represent or encode cognitive processes in the vertebrate brain cannot be fulfilled with current technology or concepts. This does not mean, however, that there will not be wonderful and amazing progress in determining how individual neurons or particular parts of the brain may be involved in particular activities or unraveling the mysteries of the anatomy and physiology. The extraordinary new work of Nabavi and his colleagues being a wonderful example of how we may be able to learn some things (e.g., the general proposition that changes in synaptic conductivity encode memories) while others (e.g., the specific synaptic pattern differences that encode different memories) remain obscure. There are many other examples of what would have seemed to be impossible only a few years ago; the accomplishments in understanding the ionic forces that explain how individual neurons work and interact at the cellular and molecular levels are among the older crown jewels of neurophysiology. So, too, are the gross anatomical studies of the macroconnections within the brain that have been made possible by the development of devices such as the diffusion tensor MRI (Jones, Simmons, Williams, and Horsfield, 1999). However, the specific mapping of the microneuronal networks that most cognitive neuroscientists would argue logically are the "psychoneural equivalents" of cognition remain elusive, if not absolutely inaccessible.

In this chapter, I have concentrated on the role of networks in current thinking, both in terms of available technology and our conceptual underpinnings. The most advanced technological development that may be of use in studying networks has been progress in multiple electrodes arrays. Unfortunately, there remain many obstacles to the use of arrays of MEAs to record or stimulate the activity of many neurons. These include:

1. Microelectrode arrays of this kind characteristically consist of electrode tips that are larger (60 μm) than single microelectrodes (1 μm). Therefore, they are limited to recording only extracellular responses.
2. There are two tasks to be accomplished for which the MEA initially seemed well suited: (1) the assignment of recorded responses to their neuron of origin, that is, the clustering or sorting problem, and (2) the determination of the neuronal map indicating which neurons are involved and what are their interconnections. However, there are serious impediments to accomplishing either of these analytical tasks.
3. The amplitude and other properties of recorded neuronal responses will vary depending on the fortuitous spatial relationships between neurons and electrodes—relationships that cannot be predicted in advance.
4. It is, therefore, extremely difficult to definitively determine which responses are coming from which neurons with current technology. Different neurons

may display the same shapes and amplitudes and the same neuron may be characteristically different from one trial to the next. No single discriminating criterion for solving the clustering problem works in a general way.

5. Recording is fortuitous: Some electrodes will record nothing and some will pick up the responses from multiple neurons in an unpredictable manner.
6. The regular pattern of an array of 100 such microelectrodes (e.g., as shown in Figure 3.2B) does not necessarily correspond to the spatial arrangement of the neurons in the neural tissue under investigation. That is, beyond the sensory and motor representations, they may not be topologically consistent. Therefore, it may not be possible to map directly from electrode recordings to the irregular pattern of a cognitively significant neuronal network.
7. The irregular distribution of synaptic connection also contributes to the lack of repeatability of experiments.
8. The biological and chemical responses to the foreign matter from which the electrodes are constructed can mitigate their use in long-term applications—a requirement for clinical use. Similarly, long-term implantation of electrodes remains a major challenge. A good discussion of the difficulties encountered when one attempts to record or stimulate for long periods can be found in the review by Moxon et al. (2009).
9. Even the most minute array of electrodes will do damage to neurons and interconnections. Efforts to develop flexible and compliant microelectrode arrays have met with only moderate success.
10. Much more devastating to the hope that the organization of realistic neural networks will be uncovered using MEAs is the enormous amount of data that is obtained from even a relatively few irregularly spaced neurons. If the goal is to understand how multiple neurons interact, then the persistent problem of combinatorial computability produced by the complexity of neuronal networks cannot be overlooked.

The thesis of this chapter is that partly because of technical and procedural limitations, but mostly because of the intrinsic nature (i.e., the complexity) of the problem, the direct, exhaustive, brute force strategy of stimulating and recording the individual components of a microneuronal network that worked so well at the level of single neurons is not likely to ever be implemented at the microneuronal level for ensemble processes. Microneuronal network theory building in such a many-neuron environment is severely inhibited because of the improbability of ever simultaneously and controllably stimulating and measuring the activity of the number of neurons involved in even a "simple" cognitive process. Thus, the possibility of developing an authentic, valid, overarching microneuronal theory of cognition based on robust empirical evidence appears to be remote. This is so despite the fact that there are no plausible alternatives to Hebb's conjecture and the speculative theories that flow from it. Hebb was right, most cognitive neuroscientists seem to agree, but there is no way to prove it in detail. Although we may all agree that, in principle, cognition is encoded by synaptic mechanisms, the specific relation of mental processes to the detailed states of the microneuronal network that represent our thoughts is likely to remain permanently elusive.

In conclusion, despite some remarkable technical accomplishments, such as development of MEAs, technology has not yet begun to contribute answers to the great question—how do neural networks encode or represent cognitive processes? There appear to be fundamental limits to what can be accomplished with this type of multiple electrode device. Difficulties, including the lack of selectivity of the individual electrodes of an MEA, the irregularity of neuronal network organization, and the usual problems with poor signal-to-noise ratios will perpetually obstruct progress toward a comprehensive neuronal network theory. Many other technical difficulties exist, some of which may be solved in the future whereas others will remain to perpetually frustrate the desire to study the action of microneuronal networks. Looming over everything are the problems of computational explosions and data storage and manipulation because of the complexity of neuronal networks.

An additional conclusion to be drawn from this discussion is that naïve hopes of building a brain with as many synthetic "neuromimes" is probably also impossible. Not just because we do not have suitably sized supercomputers (we are close to that goal) but because the logistics of stimulating and recording from them are beyond us. These huge numbers of real neurons speak directly to any hopes that we might have had for general or direct computational demonstrations in support of Hebb's conjecture and, as noted, restrict us to theories and experiments involving only a few neurons—abstractions known as "toy networks." This pessimistic, but realistic, argument is sound because it is almost certainly true given that immense numbers of neurons are involved in even the simplest cognitive process.

On the immediate scene, complexity as we have presented it here has profound practical effects. First, those who toil in the laboratory seeking the neural codes underlying cognitive processes are inevitably and unavoidably going to be disappointed barring some extraordinary reconceptualization of the problem. Second (as suggested to me in a personal communication from the Arizona State University neurophysiologist Steven Helms Tillery), many originally optimistic newcomers to the field have now begun to withdraw from efforts to map neuronal network topologies except in the context of sensory and motor systems. These peripheral transmission systems may help us to understand certain general principles of network organization but cannot penetrate the specific codes for cognitive processes.

A concise summary of the review presented in this chapter is that, however plausible and unchallenged Hebb's conjecture may be, there is almost no empirical evidence to accept its basic postulate—that cognitive activity is encoded by specific patterns of activity in microneuronal networks. Despite much published speculation revolving around this question and the development of ingenious

TABLE 3.1 Relationships between Theories and Technologies

Theories	*Technologies*
Mind is in the heart	Dissection and trauma
Mind is in the head	Behavioral observations
Mind is distributed	EEG
Mind is localized	fMRI
Single neuron theory	Microelectrodes
Neuronal networks	Computers and MEAs
Hebb's conjecture	Plausible speculation

Source unknown.

devices and complex analytical techniques, the goal of mapping a cognitive process onto a neuronal network remains an unfulfilled hope. Experiments may optimistically start with MEAs and elaborate statistical techniques, but they quickly founder in a sea of complex interactions and numbers far too large to be processed by even the most powerful computers of which we can conceive. Although success with the relatively simple task of sorting SAPs provides an illusion of success, the gap between spike sorting and mapping the neuronal equivalent of a thought remains unbridged.

If this view of the neuroreductionist approach to cognition turns out to be correct and realistic, albeit pessimistic, how, then, can we account for the continued effort to try to crack a code that may be uncrackable? The answer to this query must lie in one of the premises of this book—namely that both our experimental protocols and our prevailing theories of mind–brain relationships are in large part determined by the technology that is available rather than robust empirical findings. To highlight this relationship, consider Table 3.1.

In each case, theory is driven by the kind of instruments available, producing a particular kind of data. The major exception to this generality is that the most widely accepted theory—Hebb's conjecture—is also the least substantiated by solid evidence.

Notes

1. The optogenetic technique is a major step forward in neurophysiological research. By various methods (such as using a virus that carries a light-sensitive trait) genetic material can be introduced into a neuron (or, more usually, a class of neurons) that makes it respond differentially to light stimuli of different wavelengths or temporal patterns. Manipulations of opsins, the light-sensitive materials of the eye, are particularly useful in this regard since their optical properties are well known. Because of the different optical sensitivities of the genetically manipulated materials, specific neuronal or synaptic types can be turned on or off by stimulating the neurons with different kinds of light. In the Nabavi et al. study the virus used to vary the behavior of the synapses produced neurons that were selectively sensitive to low-frequency trains of light. Specifically, the optical

stimulus–producing LTD was 900 pulses of light, each 2 msec, at 1HZ (Nabavi et al., 2014). The stimuli-producing LTP was 5 trains of light (each train 100 pulses, 100Hz. In this manner, the degree of synaptic activity could be manipulated and direct control over their function achieved for the first time.

2. The three-body problem has been most extensively studied in physics in the context of planetary motion. It has been known since the 19th century that the behavior of three gravitational interacting bodies cannot be solved in general unless one of the bodies is much smaller than the other two. The dynamic behavior of three equal-sized objects is, as far as we know, nonrepetitive and unpredictable.
3. Sporns' (2011) book is an outstanding discussion of the nature of both anatomical connectivity and connectivity in general. It is a necessary part of the education of any cognitive neuroscientist. It tells us much about networks, but it says very little about the relation between the anatomical connectome and cognition.
4. Unfortunately, this work seems not to have been published (it is dated March 31, 2004); more complete details can be found in a single entry to it here: http://web.mit.edu/9.29/www/neville_jen/jkarr/mea/. It was, however, so interesting in the present context that I could not ignore it.
5. It was not entirely clear from their report whether they collected the data they analyzed or whether prerecorded data were provided to them for analysis. In an April 2014 personal communication from Karr, he indicated to me that they used prepackaged data as the raw material for their analysis. For the present discussion, it does not really matter since the goal of their paper (and of the present discussion) was to infer the connectivity of the microneuronal network described by the data. Having prepackaged data made the class project possible.
6. Of course, this number will change depending on whatever computer is available at any given time. However, the numbers are such that the Stockmeyer theorem argues that no conceivable computer could perform the necessary calculations in the lifetime of the universe. Whatever the specific numbers, this theorem does provide an intuitive foundation for the gap between computational needs and computational capabilities.

4

INTERMEDIATE LEVEL NEURONAL NETWORKS—THE ROLE OF MODELS

4.1 Introduction

A major goal of modern cognitive neurophysiology is to establish connections between cognitive processes and the pattern of neural activity embodied in networks of neurons. Obviously, however, this is going to be an extremely challenging goal for systems as complex as the human brain where innumerable neurons are presumably involved in the simplest of thoughts. When dealing with complex networks potentially involving millions and billions of neurons, the technical and conceptual problems inhibiting this kind of research are profound as was pointed out in the previous chapter. For all practical purposes, combinatorics and numerousness probably will not permit experiments of the necessary theoretical kind—for the present, at least, no empirical neurophysiological studies of the detailed arrangement of cognitively realistic neuronal networks have yet been carried out. An alternative strategy, of increasing importance in cognitive neuroscience, is the use of relatively simple, naturally occurring model systems that mimic or model some aspect of the behavior of much more complicated organisms—hopefully with a simple analyzable neural network. The immediate goal of this line of research is to at least understand the basic principles that may help us understand how very complex neural networks operate by studying the accessible and tractable neural networks of a simple creature.

Lying between the extremes of single neurons and full-blown neuronal networks, however, is an intermediary region in which there is a body of empirical neurophysiological evidence and for which the networks are simple enough to permit at least approximate descriptions of their functionality. This approach depends on the discovery of relatively simple neuronal networks in "model" organisms that are capable of mimicking what are almost certainly more complex analogs of vertebrate neuronal networks. Invertebrate creatures have a long history of use as simplified surrogates for vertebrate nervous system functions. I discussed one—the squid *Loligo vulgaris*—in Chapter 2's discussion of single neuron technology where anatomical freaks (e.g., large size neurons) permitted breakthroughs in our understanding of the basic physiology of neurons. This chapter takes the discussion of invertebrate model organisms that have served us well in understanding how relative small networks can be related to behavioral functions. The most important reason for the role model preparations play in current cognitive neuroscience is that they provide a mechanism for the study of neuronal networks that are otherwise too complicated to provide any hope of unraveling the network structure.

To be used as a model of a vertebrate system, a simple invertebrate system must meet certain criteria. These include the following:

1. The basic assumption underlying this kind of research is that behavior is controlled by neuronal networks—that is, interactions among a number of neurons.
2. The immediate goal of the research described here is that specific neural networks can be associated with specific behaviors.
3. The long-term goal is to understand how neuronal networks represent cognitive activity. However, in most vertebrates, the number of neurons controlling mental activity is so great as to currently pose an intractable problem. Therefore, cognitive neuroscientists have turned to invertebrate model organisms whose salient neural networks consist of only a few neurons.
4. The number of neurons in a model neuronal network necessary to induce a particular behavior must fall into an intermediate range bounded at one end by a single neuron (see Chapter 2) and at the other end by practical and computational limits on the analysis of cognitively significant neuronal networks. Typically, in this kind of research, the upper end of this range varies from a few to as many as a dozen neurons. Beyond that lies the analytical intractability typical of even slightly more complicated networks.
5. Simplifying regularities (e.g., orderly and regular networks), however, may occasionally increase the accessibility of networks consisting of as many as 1,000 neurons (see the section on the modeling of the Mach Band by the horseshoe crab *L. polyphemus* later in this chapter).
6. The model organism must have analogous behavioral properties to make it relevant to the modeler. However, these properties are not constant from preparation to preparation. Size and accessibility of the modeled neurons usually help but so, too, does a quick turnover in the life cycle of the model animal.
7. If the neurons of the model animal are large and their responses easily recorded, either intracellularly or extracellularly, this is advantageous.

8. It also helps if the nervous system of the model is stable from specimen to specimen so that the involved neurons can be individually identified and experiments be repeated.
9. Because many experimental procedures are now based on genetic manipulations, having a more complete knowledge of the model animal's genome is also advantageous.
10. The design of an experiment must be able to distinguish between a necessary and a sufficient theory of network organization (Olsen and Wilson, 2008).
11. Analogous behaviors must be carried out by both the modeled and the model system. "Analogous" in this case means that the behavior accomplishes the same function (such as learning) in both the modeled and the model organisms. It does not mean that exactly the same kinds of networks are used to implement both. Indeed, we typically do not know what mechanisms exist in the modeled organism to produce the behavior—that is why we have turned to simpler models for enlightenment.
12. The behavior of the model organism must be measurable and the network configurations controllable in order to carry out controlled experiments. A variety of methods, including genetic and neurophysiological techniques, are now available to control behavior.
13. The a priori judgment that the structure of the model is exactly the same as the structure of the modeled network is almost never provable. The relationship between the underlying mechanisms of each is one of analogy, not homology, and though behaviorally similar, there remains uncertainty about the salient networks of the model and the modeled, respectively.

If an experiment meets these criteria in a way that permits suitable experiments to be carried out, then the relationship between behavior and simple neuronal networks observed in the model can be extrapolated *by analogy* to explain comparable behavior in the modeled vertebrate. Inferences may then be drawn about the possible similarities in the underlying neuronal networks of the modeled and the model, respectively. However, it is essential to appreciate that, however robust the data, this is not proof that the kind of simple neuronal networks observed in the model exist in the real system. Modeling experiments of this kind are suggestive but not definitive. The problem is that such specific behaviors as classical conditioning could be instantiated in vastly different neuronal networks in the invertebrate model and the vertebrate system.

By judicious care in carrying out truly relevant experiments, a neuroreductionist theory of cognition may be generated. If the number of involved neurons is sufficiently small, then it is possible that a mathematical model of the observed neurophysiology can also be constructed to describe how a modest-sized network of a few neuronal components in an invertebrate can produce what is hoped is the same behavior as that produced by the millions of neurons in the vertebrate nervous system. (A comprehensive review of the application of model systems to a wide variety of fields in biological and medical research can be found in Wilson-Sanders, 2011.)

This strategy of using models to understand the neuronal networks of much more complex organisms is confounded by one great uncertainty—are the analogous behaviors exhibited by a simple neuronal network in a model invertebrate and the complex response mediated by a vertebrate organism really the same? Or are they just functional analogues mediated by different neuronal networks serving different behavioral needs? If they are just functionally similar behaviors mediated by vastly different neuronal systems, what conclusions can be drawn about the relationship between the simple responses of a model preparation and the highly complicated psychological processes of a human (e.g., in the former case, habituation of neuronal responses in *A. californica* and, in the latter case, habituation to pain in humans)?

Nevertheless, invertebrate model preparations consisting of a few neurons by virtue of their very simplicity, if nothing more, permit a kind of plausibility test of neuroreductionist theories that is not possible in a vertebrate brain. They can demonstrate that basic principles of neuronal organization, such as lateral inhibition, habituation, accommodation, adaptation, feedback, or learning (i.e., behavioral change as a function of experience), actually exist in microcosm as well as the macrocosm. Furthermore, to the degree that these functions are common to both the modeled and the model, they might well resolve some of questions of how these processes might be represented in the macrocosm of the brain. Theoretical inferences of this kind also permit the execution of mathematical analyses that are difficult or impossible to carry out in higher-level nervous systems. Thus, they can also serve as heuristics, stimulating theories that are totally intractable to analysis in the complexity of higher organisms. Not the least, they can be used as surrogates for human studies that are prohibited for ethical reasons—none of us are likely to insert electrodes or excise tissue in a human unless there is an overwhelming medical need.

It is in this spirit of a partial, incomplete, and approximate approach to the problems of neuronal representations based on model preparations and intermediate levels of coding complexity that I review in this chapter four relatively simple model systems. Studies that are based on hypothetical neuronal networks, rather than real biological observations, will receive only secondary attention. The problem is that not all intermediate scale network studies or theories are based on even simple neurophysiological data. An entire field—"neurocomputing" or "computational neuroscience" (e.g., see the work of Hertz, Krogh, and Palmer, 1991; Anderson, Pellionisz, and Rosenfeld, 1993; Ham and Kostanic, 2000)—has grown up in which simplified neuronal networks are constructed (rather than observed) in order to study the general properties of networks. Many have no real neural data supporting them. The goal of these "dry" neural network theorists is to develop the mathematical foundations of cognitively significant networks in a world of artificial neuronal networks rather than to link them to real neurophysiological systems. Exercises in neurocomputing are often based on highly simplified hypothetical neuronal networks that, more often than not, draw only limited inspiration from real biological systems. Indeed, once past the general idea of a network and parallel processing, it is rare to find an analysis of a system based on real biological findings.

Why is this true? The answer to this question lies in the nature of real neural networks. They are complicated to the point of computational intractability by their irregularity of function, idiosyncratic interconnectivity, and sheer numerousness. The result of these intrinsic impediments to dealing with real biology has been to shift the emphasis from the study of understanding biological networks to the pursuit of strictly engineering topics (e.g., parallel processing) although retaining the "neurocomputing" nomenclature. Some of the most significant progress in the field of computational neuroscience has, thus, come from research that was only distantly related to real neural networks. The major works of such notable pioneers in the field of neurocomputing as McCulloch and Pitts (1943), Rosenblatt (1958), and Hopfield (1982) were all carried out with only perfunctory allusion to their original biological influences. These distinguished pioneers often simplified their models to achieve mathematical tractability to the point of irrelevance to biology as they made efforts to explain how simplified networks might plausibly imitate biological functions.

Another issue should be considered at this point. The importance of model systems is indisputable. However, there is an argument that even the advantageous opportunities they represent may have the ability to misdirect our research and theoretical activities in the same way that technology does. Just as the availability of a measuring instrument may influence the kind of thinking that goes into planning a research program, so, too, do the accidents of the availability of a particular model system guide and constrain our thinking about the neuronal substrates of a behavior. It seems quite clear, for example, that despite the fact that it seems to provide a compelling foundation for a theory of perceived edge enhancements, *L. polyphemus* did not evolve as a convenience for Hartline's group. It was available and an analysis of its network properties was sufficient to analogize a particular visual phenomenon. However, proving its unique necessity as an explanation for the perceptual phenomenon of edge enhancement reported by human observers requires some strategies that may not be possible—for example, the manipulation of the spatial stimulus and recording from many neurons simultaneously. This may be considered to be a quibble, but the accident of availability will always have to be a consideration—particularly when choosing between competing ideas. Proving sufficiency is easy; proving the necessity of a particular configuration is far more difficult.

The goal of the rest of this chapter is to show how a sampling of model invertebrate systems allows us to examine real neurophysiological networks in a way that permits us to provide rudimentary theories of well-defined cognitive processes. The work I shall discuss collectively defines an intermediate level of analysis in which actual neuronal networks are simple enough to be associated with some well-defined behavior. We start with what is both a classical study and a persistent example of perhaps the most successful and comprehensive neuronal model of a well-known perceptual phenomenon yet developed—the edge enhancement illusion known as the Mach Band emerging as modeled by the horseshoe crab.

4.2 *Limulus polyphemus* and the Mach Band

Since the time[1] of the great physicist, philosopher, and early cognitive neuroscientist Ernst Mach (1838–1916), scientists have known that edges between two different luminous regions of different intensity would be perceived as being enhanced on both sides of the edge. That is, the contour on the side of the brighter region would appear to an observer to be brighter than simple radiometric measurements suggested it should be; and the contour on the dimmer side would appear to be darker than the same physical measurement would suggest it should be. Figure 4.1 shows the physical stimulus distribution and the perceptual experience. These perceived edge enhancements are illusory; they cannot be detected by the most precise measurements of physical stimulus intensity and, therefore, must be the products of neuronal transformations carried out in the nervous system.

As long ago as 1865, Mach presciently appreciated that this phenomenon was associated with some kind of interaction between different points in the visual scene. He was probably the first to have stated that the Mach Band was produced

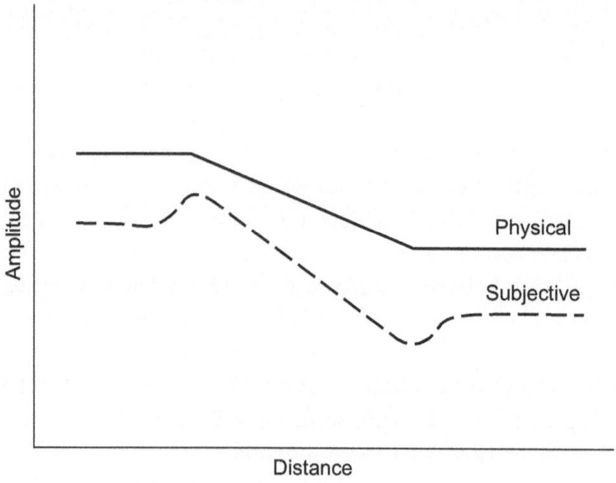

FIGURE 4.1 The shape of the physical and subjective responses to a gradient of illumination.

Reproduced from Uttal (1973).

by reciprocal interactions between different points on the retina and/or the rest of the perceptual nervous system. Of course, in the 19th century, it was not possible for researchers to carry out the necessary experiments to confirm this hypothesis, but in the latter half of the 20th century, it became eminently feasible, if not yet on the vertebrate visual system, then on a simpler model preparation—the visual system of *Limulus polyphemus*—the North American horseshoe crab.

The horseshoe crab is a vestige of organic evolution that has survived from roughly 400 million years ago. It is an arthropod but actually unrelated to crabs and other crustaceans; it is more closely related to spiders and other arachnids. Fortunately for cognitive neuroscience it is also graced, like spiders, with multiple eyes on the top of its carapace. Two large compound eyes, in particular, are of special interest because each consists of approximately 1,024 relatively large cell complexes called ommatidia. Each ommatidium contains a number of secondary cells, but lateral neuronal connectives from the eccentric cell (the main light-sensitive neuron in the ommatidium) are the main links to other cells of the complex ommatidium. By this means, light-induced neural responses interact with surrounding ommatidia through lateral connections that can be seen in optical microscopes (as shown in Figure 4.2—a drawing of the neural network in one ommatidium).

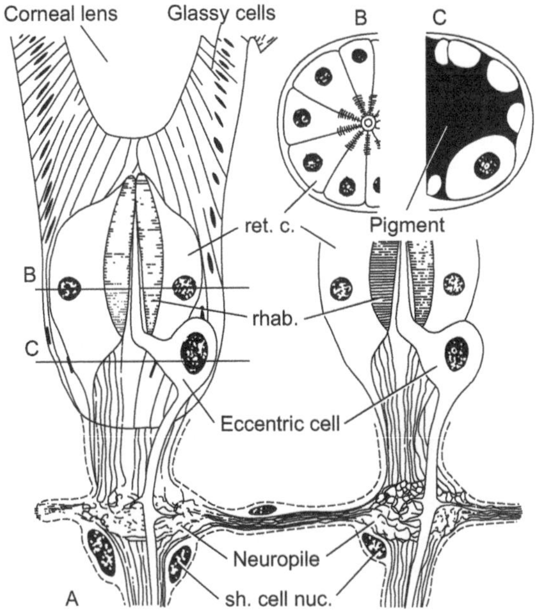

FIGURE 4.2 The anatomy of an ommatidium in the *L. polyphemus* compound eye. Analogs of the lateral connections at the base of the figure are believed to account for the Mach Bands.

Reproduced from MacNichol (1956) with permission.

The relatively huge size of the cells in the ommatidia first led researchers (Hartline and Graham, 1932) to use them as a model preparation for the study of single neurons. However, an inspired epiphany stimulated the group led by Hartline (1940) to consider the *L. polyphemus* eye as a model of interneuronal interactions. The basic paradigm that Hartline and his coworkers used was to stimulate a narrow region of the crab's eye with a spot of light, measure its neural activity with electrodes, and then stimulate it with another spot in a different location. The typical result was that the addition of the second light stimulus would reduce the response from the ommatidium from which the recording was made. Although only a single electrode was used to make this measurement (and the response of only a single ommatidium recorded at a time), both the electrode and the two spots of stimulating light could be moved across the compound eye to track out the spatial pattern of the response as a function of such variables as the distance between the two light spots. By this reiterative procedure, it was thus possible to deal with this model preparation as if multiple electrodes were recording from a spatially distributed response to a spatially distributed stimulus—sequential recordings taking the place of a simultaneous spatial distribution of responses.

This technique may have been the first experiment to deal neurophysiologically with neurons interacting in a microscopically defined real neuronal network. Previously only recordings had been made from single neurons. The work carried out by the Hartline group was, therefore, a major milestone in the study of neuronal network theory thanks to the unusual nature of an evolutionary anomaly—*L. polyphemus*. Although the number of neurons in the creature's eye was larger than what I have designated here as an intermediate level, the extreme regularity of the anatomy made it possible to deal experimentally with the network properties of this unusual eye. It is problematic whether this feat has been repeated in any other model preparation.

The specific question asked in this pioneering series of experiments was—does the presence of the second stimulus affect the first region's response (Hartline, 1949)? The answer to this questions was that yes, it certainty does and in a very specific manner whose properties could be measured—the sine qua non of a model preparation. The interactions between ommatidia were solely inhibitory, diminished in strength with the distance between stimulated regions, and were also mutual (Hartline, Wagner, and Ratliff, 1956)—the effect of one on another was the same as the other's on the first. The main implication of this research was to open the door to the study of the simultaneous interactions of many neurons in the form of a network. It provided a compelling explanation of the Mach Band phenomenon by demonstrating a neurophysiological analog of edge enhancement, thus validating what had only been an inspired guess by Mach.

This simple model invertebrate system thus also provided a compelling explanation of the similar neural interactions that presumably were going on in the human visual system. It was determined that these lateral inhibitory interconnections in the eye of *L. polyphemus* produce contour enhancements in the animal's compound eye that were at least analogous to the psychophysically measured Mach Bands in humans, as shown in Figure 4.1, if not homologous. Because of the regularity and simplicity of the neural mechanisms, it immediately became obvious that a formal mathematical model of the system was possible. The formularization of the appropriate mathematical model was carried out by Hartline and Ratliff (1958) in the form of a system of simultaneous equations, shown in Equation 4.1. Each of the equations in this system calculated the response (r_p) of each ommatidium as positively influenced by the stimulus excitation (e_p) and negatively by the sum of all inhibitory influences from its neighboring ommatidia. In Equation 4.1 the sum of the inhibitory forces from n cells is modified by a constant (k_{pj}) specific to the interaction between any pair of ommatidia. The term ($r_j - r_{p,j}$) refers to the threshold of response for each ommatidium:

$$r_p = e_p - \sum_{j=1}^{n} k_{pj}\left(r_j - r_{pj}\right) \qquad \text{Eqn. 4.1}$$

There are several important aspects of Hartline and Ratliff's mathematical model, not the least of which is the fact that it was one of the first such theories to take real neurophysiological data (not just some speculations about hypothetical mechanisms) and express them in a form that could be numerically evaluated. In many ways it was nearly an ideal model preparation. The number of ommatidia was small (1,024), they were very large, and they were interconnected in virtually the simplest possible manner—mutual lateral inhibitory connections. The arrangement of these visual units of the *L. polyphemus* eye was also highly regular; indeed, the orderly arrangement of the ommatidia is almost crystalline in the way they are distributed and interconnected within the compound eye. Thus, it was possible to manipulate a real neuronal network by varying some of the key properties in order to determine some of the values that characterize the real system but might not be directly measurable neurophysiologically.

Most important of all is that this model preparation responds in a way that is comparable to the behavior of the system being modeled. Comparable in this case means that it—the model—follows the same laws and tracks the same trajectory that the real world process does producing the same phenomenological outcome. In other words, the model's and the real system's behavior are functional analogs of each other. In short, they are doing the same thing—enhancing edges that are not present in the physical stimulus! But are they really the same?

This question brings us back to the caveat expressed earlier—is model system research producing the same behavior (i.e., edge enhancement) because of the same underlying mechanism—lateral inhibitory interaction—or are the common behaviors simply analogous response outcomes of what are actually different mechanisms? The point is that it is entirely possible that there is more than the one explanatory model on which this explanation may be based. There is no way to determine if this is the case in the *L. polyphemus*–human comparison situation because there is no way to determine if the human perceptual Mach Band is caused by the same kind of lateral inhibitory interaction used by *L. polyphemus*' visual systems. Our ability to use the model system to examine the actual neurophysiology of the network in *L. polyphemus* is not matched by our ability to examine human neurophysiological processes. There are alternative possible systems that involve non-lateral inhibitory interaction, perhaps in the form of symbolic[2] rather than the geometrical interactions epitomized by Mach's famous theory. That alternative explanations may exist as plausible, if undemonstrated, alternatives is suggested by some discrepancies between the way *L. polyphemus* produces the neurophysiological analogs and the nature of related human illusions. Specifically, I refer to the equally well-known *simultaneous contrast* phenomenon, which distributes the enhancement effects equally over the entire spatial extent of an illuminated region surrounded by a light of different intensity rather than just at the edges; there is no edge enhancement in this illusion yet it is typically "explained" by the same lateral inhibitory interaction mechanism as the Mach Band. Both cannot be correct.

What does result from Hartline's analysis is, therefore, a highly plausible theoretical network description of a complex visual process based on certain functional analogies and robust neurophysiological data. Given the current state of knowledge of the origins of our perceptual experience it, therefore, becomes defensible, if not in a definitive way, then in a highly plausible and possible way to say that we have "explained" the Mach Band phenomenon.

The advantage of using *L. polyphemus* as a model of a human perceptual phenomenon is that the anatomy and physiology of the model animal is so simple and stable from one animal to another that it can be used to study one possible neural explanation of the perceived Mach Band in a direct way that would not be amenable to direct experimentation in humans. To have such compelling forms of analogous behavior in both the recordings from *L. polyphemus* and the spatial dimensions of perceptual experience reported by humans was a major part of the success of this project. The specific parameters of the mathematical model are based on specific anatomical and neurophysiological measures obtained from experimentation. No a priori assumptions about the underlying mechanisms were necessary although in this simple system, a genius such as Mach was able to presciently infer an explanation based on reciprocal inhibitory neural interactions de novo. In more complex networks, preliminary assumptions may be misleading and the possibility of alternative explanations in the form of other kinds of brain functions cannot be completely rejected. For example, the illusion may also be explained in terms of the high band pass properties known to exist in the visual pathway—an entirely different process than the reciprocal lateral inhibitory interaction on which the Hartline model is based. None of this diminishes one of the great success stories in cognitive neuroscience.

This does not mean that the final explanation of this phenomenon in humans is at hand. That the neurophysiological properties underlying edge enhancement in the *L. polyphemus* and the human perceptual phenomenon are mediated by exactly the same simple neuronal networks seems intuitively unlikely. The Hartline-Ratliff "theory" that the neural and perceptual phenomena are alike in principle is based on a very simple sensory mechanism for which there is ample neurophysiological supporting data and in which the neuronal network is relatively simple in structure. In other words, although the *L. polyphemus* eye effect is based on local geometric interactions, we still have not established that this is also the case in the human. Nevertheless, what we do know about the neurophysiology and neuroanatomy of *L. polyphemus* is probably still the best model we have today in which the neural and psychophysical data are so mutually consistent.

In sum, the Hartline et al. research program was one of the first, if not the first, to develop a mathematical theory of observed neurophysiological network mechanisms that might underlay a psychophysically reported illusion—a discrepancy between the stimulus and the resulting experience based on the properties of an interactive network. It did so by providing a body of relevant experimental data that built on direct measurements of neurophysiological responses. The model has been widely accepted because it reproduced the perceptual experience in a simple model neurophysiological preparation. It is however, based on another unproven assumption, namely that the spatial mapping of the neuronal response in *L. polyphemus* is recapitulated in the perceptual experience of the human observer. This gap in the logical chain remains unbridged.

The key to this theoretical success was the susceptibility to mathematical modeling provided by the simplicity of the *L. polyphemus* eye. An important aspect that makes this analysis so compelling is that the mechanisms used to explain the overall system responses are from microneuronal interactions—that is, network properties. The edge enhancement phenomenon would not be observed in the activity of a single ommatidium. It was only when multiple components were shown to interact with one another that the neuronal connections and interactions of this compound eye exhibited the edge enhancement phenomenon. The presumption is that similar (in principle) but probably much more complex neuronal networks account for the Mach Band perceptual illusion reported by humans. If nothing else, this work was one of the earliest to provide an empirical foundation for a fundamental postulate of cognitive neuroscience—that it is from the information-processing capabilities of neuronal networks that a preliminary understanding of cognition is most likely to come.

4.3 *Aplysia californica* and Nonassociative Learning

If *L. polyphemus* provided the first and still most famous vehicle for the development of an intermediate level network theory of sensory or perceptual function, it was another quite different animal that provided a model for a kind of behavioral change referred to as Nonassociative Learning (NAL). NAL is a change in behavior caused by simple experience with stimuli. Unlike classical or instrumental learning in which some additional stimulus must be associated with the one under study to reinforce or modify behavior, NAL depends purely on the number of times a stimulus is experienced. Thus, an experiment studying NAL may involve nothing more than simple repetition of a stimulus and measurement of changes in a neuronal response.

It is well established that NAL occurs in humans—engaging in repetitious tasks leads to a reduction in performance. Furthermore, we learn not to be aware of the constant stimuli coming from our clothing or the repetitious ringing of a bell in the background of our environment. In general, this reduction in response strength resulting from a repetitive or constant stimulus is a form of NAL behavior called habituation. Habituation is a relatively easy protocol to study experimentally. All one has to do is repeatedly stimulate a preparation and observe the strength of the evoked response.

A related phenomenon is the rapid recovery from habituation referred to as dishabituation. Dishabituation occurs when the quality or locale of a stimulus changes, that is, when a novel stimulus is presented. Habituation and dishabituation are supposedly the result of a reduction in the behavioral salience of the stimulus but until recently, where in a neuronal network this kind of learning resided could only be speculative. The behavioral significance of habituation is that repeated stimulation that does not harm the animal tends to diminish in behavioral salience. Habitation is thus neatly and concisely defined as a response decrement as a result of repeated stimuli that is not due to sensory or motor fatigue.

Experimentally distinguishing habituation from sensory or motor fatigue is not easy; both processes represent decrements in performance as a result of stimulus repetition. For habituation, the best indicator is that after habituation, when a novel or a nonspecific stimulus is presented, the original response *immediately* recurs with a strength equal or greater to that obtained before the habituation stimulus was initially presented. To the contrary, if the response is due to fatigue, the response produced by a novel stimulus would still be attenuated and would remain so until the sensory or motor mechanism recovered from whatever resource exhaustion had diminished its response strength. In sum, habituation is a decrease in responsiveness with repeated stimuli not due to fatigue. Other suggested criteria for distinguishing habituation from fatigue, such as the exponential shape of the decrease of the response as a stimulus is repeated, are not definitive since fatigue and habituation generally produce similarly shaped response decrement curves.

A closely related, but inverse, phenomenon to habituation is sensitization—an *increase* in response strength with repeated stimuli. Sensitization is considered to be a defensive response to stimuli that are noxious. Sensitization, however well documented, is more elusive of definition, but to use a parallel with disinhibition, we may define sensitization as a response increment as a result of repeated stimuli.

Because of the conceptual simplicity of NAL as a highly reduced form of learning, explorations of its neuronal origins raised the possibility that an animal model of habituation might illuminate some of the basic principles of learning in general. Gerard and Forbes (1928), for example, in an early article, had studied what they called "fatigue" in decerebrate cats. It now seems that their methods suggest that the response they observed may have been better defined as a form of habituation. The state of psychological studies of NAL has been periodically reviewed by investigators such as Harris (1943) and Thompson and Spencer (1966) and, although currently less of an active research topic than previously, its simplicity as a neurophysiological model of learning has rejuvenated interest in the field.

In the 1960s and 1970s a team led by Eric Kandel, who was subsequently to receive the Nobel Prize for this work, began to study habituation and dishabituation in the sea hare (*Aplysia californica*, a gastropod mollusk) as a model of simple forms of learning. *A. californica* has some highly desirable features that made it a superb model preparation. It has a relatively simple nervous system and yet has been shown to exhibit a variety of different kinds of learning, including both NAL and associative types of conditioning. The neurons in its nervous system are large and relatively few—approximately 20,000. Furthermore, many of the neurons are both individually identifiable and constant in position and shape from one animal to the next.

A. californica's natural behavior included two defensive reflexes to protect the sensitive tissue of its siphon and gills. These reflexes are mediated by different neurons in the abdominal ganglion and elicited by what turned out to be very simple neuronal networks. Some motor neurons (e.g., LD_{s1}) in this ganglion controlled contraction of the siphon while others (e.g., L7) controlled gill retractions when stimulated by what could be dangerous or toxic stimuli or even slight touch. Because of what turned out to be a behavioral similarity to human NAL and the relative simplicity of the neuronal network controlling this behavior, *A. californica* became a model animal for the study of habituation and dishabituation in humans. This relevance could be enhanced because the animal proved to be robust enough to have the germane portions of its sensory and motor systems surgically isolatable, further simplifying the model system down to a minimum number of neurons and synapses—a number that surprisingly turned out to be mediated by a single synapse.

The early experimental designs to study NAL in *A. californica* were straightforward. Stimuli were typically mechanical (droplets of water) although in some other work flashes of light or small electrical stimuli were used. Motor responses were measured in various experiments by examining neuronal responses with both intracellular and extracellular electrodes or by photocells placed in the pathway of the retracted tissues to study motor functions such as gill withdrawal. The system proved to be so simple that both electrophysiological and motor responses could be measured and compared in identified neurons.

The pioneering experimental paradigm used by Kandel and his colleagues to study habituation and dishabituation in *A. californica* was described in what are now considered to be three classical papers (Castellucci, Pinsker, Kupfermann, and Kandel, 1970; Kupfermann, Castellucci, Pinsker, and Kandel, 1970; Pinsker, Kupfermann, Castellucci, and Kandel, 1970). The first of these three reports (Pinsker et al., 1970) documented a purely behavioral study in which an immobilized but intact *A. californica* was stimulated with a "brief jet of sea water" delivered to a sensitive spot on the animal's skin. The response measured was a physical retraction (gill withdrawal) and was measured with a photocell that was tucked under the gill. Repetitive stimulation with a series of constant-amplitude stimuli showed a progressive decrement in the magnitude of the gill withdrawal response as the stimulus repeated, that is, habituation. Even after the *A. californica*'s gill withdrawal response was completely habituated, the response magnitude would promptly recover to prehabituation levels if a relatively strong stimulus was delivered to another sensitive region of the skin, that is, dishabituation. This observation and the presence of spontaneous gill withdrawals at a high amplitude level during the experiment established that the responses were not due to sensory or motor fatigue, but could be considered to be a legitimate form of NAL.

Based on this behavioral study, the next two articles in this series were aimed at determining the underlying neuronal mechanisms of this simple form of learning by establishing the neural network that mediated the gill withdrawal behavior. In the second article in this series (Kupfermann et al., 1970), intracellular microelectrodes were inserted into neurons of the abdominal ganglion. Because of their large size, individually identified neurons could be examined from one animal to the next. The stimuli used to activate the gill withdrawal response in this second study were both electrical and tactile, the former of which were delivered through a double-barreled microelectrode.

Because of the large size and anatomical repeatability of the neurons in the abdominal ganglion, Kupfermann and his colleagues were able to identify additional motor neurons that individually could produce the gill withdrawal response to stimuli. They were also able to determine that there were no changes in the amplitude of the sensory neuron response or in the neuromuscular junctions that activated the muscles controlling the gill withdrawal during the repeated series of habituating stimuli. The main surprising result of their experiment was that the electrophysiological changes mediating the gill withdrawal reflex seemed to occur in the single synapse between a sensory and a motor neuron. The neural network they developed of the *A. californica* gill withdrawal is shown in Figure 4.3. This is not an abstraction or a diagrammatic representation but a more or less complete network made up of the actual neurons and synapses involved in *A. californica*'s NAL.

The final article in this series (Castellucci et al., 1970) carried the Kandel group's analysis of *A. californica* habituation to the next level. In the two earlier papers, the gill and siphon withdrawal reflexes were examined first by behavioral

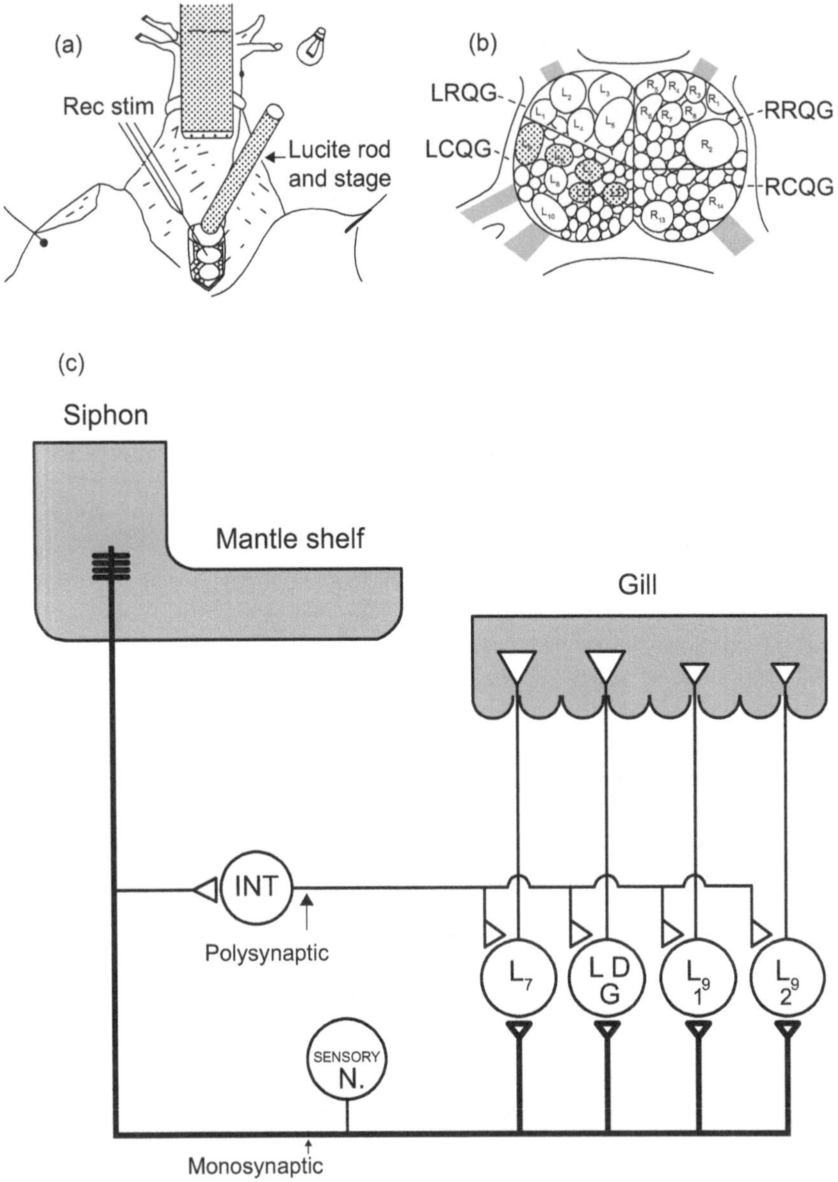

FIGURE 4.3 The anatomy of the gill and siphon neural mechanisms in *Aplysia californica.* (a) and (b) show the gross anatomy of the system, and (c) shows the six neurons involved in withdrawal from noxious stimuli, some of which are redundant.

Reproduced from Kupfermann et al. (1970) with permission.

studies and then by electrophysiological studies on the relatively intact animal, respectively. Because the anatomy of the *A. californica* nervous system was not completely known, there still remained the possibility that more complicated, but unknown, networks were redundantly involved in habituation of the response. Recall that at least four motor neurons had been demonstrated that could elicit the withdrawal reflex. Furthermore, it was subsequently determined (Zecevic et al., 1989) that as many as 300 neurons might actually be involved in the gill withdrawal reflex.

What was needed was an even more reduced model to determine exactly what was happening in this simple nervous system. The goal was precisely to determine the locus of the changes that produced habituation by eliminating other possibilities, and what better way to accomplish this than by dissecting away other possible neuronal mechanisms. To accomplish this, Castellucci et al. produced a dissected preparation that was reduced to its most fundamental simplicity—a minimal set of components that could sustain the habituation process. This minimal preparation included a patch of skin from the sensitive receptive field (the stimulation of which produced the gill withdrawal response), a single nerve "strand" passing centrally from this sensitive region, and an abdominal ganglion. In this highly reduced preparation, electrical or tactile stimuli applied to the sensitive patch of excised skin would produce synaptic responses in the motor neurons when stimulated with an impaled microelectrode. (The motor neuron responses were monitored as a surrogate for the gill withdrawal response as determined in the first two experiments.) Although there were several motor neurons, 85% of the withdrawal response was mediated by a single one—LDG1—and much of their subsequent research utilized an isolated segment of this neuron. The responses of LDG1 in the abdominal ganglion were measured as the series of the habituating stimuli were presented.

After carrying out a number of control experiments, Castellucci and his colleagues were able to show that both the habituation and dishabituation processes seemed to result from changes in the conductivity of the single synapse interposed between the sensory and motor neurons. As the stimulus was repeated, the synaptic conductivity was reduced and the response in the motor neurons progressively decreased. Similarly, when a dishabituation stimulus was applied, the synaptic response increased to its original level. Later on, Kandel's group (Cohen, Kaplan, Kandel, and Hawkins, 1997) showed that the synaptic effects were different for habituation, on the one hand, and dishabituation and sensitization, on the other. This follow-up work suggested that habituation might be produced by reduction in the sensory synapses, whereas the other two behaviors might be due to more complex networks involving such additional mechanisms as excitation of neuromuscular junctions.

In short, this research showed that habituation of the gill withdrawal reflex occurred in the simplest of possible neuronal networks—a sensory and a motor component as regulated by modifications in the conductivity of a single (monosynaptic) junction whose postsynaptic properties mirrored the behavior of the intact animal. Fatigue was ruled out in both the motor and sensory mechanisms. Whatever the details of the specific mechanisms involved, the fact that this reduced form of NAL could be attributed to a specific synaptic mechanism was a major breakthrough. A form of learning—NAL—was shown to be explained (in a very strong sense of the word) by synaptic conductivity changes. In other words, the relation between a stimulus and a behavior was *caused* by a very specific change in the properties of this simple neuronal network.

This was an extraordinarily important finding despite the fact that it came from a very simple model preparation. For what was perhaps the first time, a well-known form of learning, however primitive this form of NAL may be, could be attributed directly to synaptic changes. Although there had been much speculation (see the discussion of Hebb's prescient ideas in Chapter 3) that such synaptic changes lay at the foundation of learning, this may have been the first time that such a strong correlation between a behavioral change and changes in synaptic conductivity had been demonstrated. Of course, this particular mechanism for habituation and dishabituation—modifications in a single synaptic junction between sensory and motor neurons—might be present only in such simple invertebrates as *A. californica*; more complex networks may be necessary to explain analogous processes for habituation, dishabituation, and sensitization in higher animals. Nevertheless, the principle that changes in synaptic conductivity might account for behavioral changes reverberates down through all of modern cognitive neuroscience.

Kandel and his colleagues went on to elaborate the properties of this model preparation. Carew, Pinsker, and Kandel (1972) demonstrated that the habituation process might have very long-term effects lasting as long as three weeks, raising the possibility that not only short-term memory (in the form of NAL) could be explained, but perhaps also, a longer-term form of memory. Castellucci and Kandel (1976) subsequently showed that habituation results from synaptic conductivity reductions, sensitization is likely to be associated with presynaptic facilitation, but dishabituation is more likely mediated by post-synaptic changes. An important expansion of their work was reported by Carew, Hawkins, and Kandel (1983) in which they showed that classical as well as Nonassociative Learning could be demonstrated in *A. californica.* Nevertheless, it was the initial trio of publications that set the stage for decades of work to follow.

The most important aspect of their contribution was the confirmation of the idea that the changes in synaptic conductivity could be associated with changes in molar behavior. It must also be appreciated, however, that this example of how a model animal might trigger theoretical insights into more complex learning processes is made possible by the fact that the relevant portion of *A. californica*'s nervous system is virtually as simple as it can be in terms of the neuronal interconnections. In *A. californica*, only a single sensory neuron and a single motor neuron interconnected by a single synapse are sufficient to provide a compelling explanation of simple behavioral changes. As usual, the standard caveat must be expressed; it is not likely that the simple monosynaptic junction to which they attribute NAL in this model animal is the same neuronal mechanism as that mediating NAL in higher vertebrates. The gill withdrawal response and human habituation are at best analogs of each other. The utter simplicity of the neuronal interconnections in this preparation make any network theory trivial—the *A. californica* neuronal mechanism for NAL could be modeled at the synapse level simply by assigning graded valences to the operation of a single synapse as information is passed from a single sensory neuron to a single motor neuron. As a result, based as it is on such robust findings and relatively simple recording techniques, the simple *A. californica* nervous system has become an excellent choice for studying the biochemistry of synaptic functions, especially the role of certain ions in distinguishing between pre- and postsynaptic changes in synaptic conductivity. Research on this model preparation has mainly shifted from the nearly complete study of the synaptic information processing to the biochemistry that can explain the macromolecular changes in the synapse. For example, Gingrich and Byrne (1985) and Gingrich, Baxter, and Byrne (1988) explored the role of calcium and potassium ions in regulating presynaptic facilitation. An even more up-to-date simulation by Zhang et al. (2012) dealing with the enhancement of long-term learning protocols in *A. californica* concentrates on the role of such chemicals as protein kinase A and signal-related kinase in responding to the replication rate of stimuli in prolonging long-term effects.

The point in considering these few examples of what is certainly a large research literature is to illustrate the value of a highly simplified model preparation not only as a model of learning but also to the degree it opens the door to understanding the molecular details of synaptic action. What we must tentatively conclude from this tale of model nervous systems and behavioral changes is that it is more likely than not that synaptic plasticity is the neurophysiological basis of NAL and probably all other kinds of learning.

4.4 *Caenorhabditis elegans* and Learned Chemoaversion

Although it may be stretching the meaning of the term "intermediate level of analysis," no discussion of model nervous systems would be complete without considering current knowledge concerning the detailed system organization and anatomy of the microscopic (its body is approximately 1 mm long) nematode *Caenorhabditis elegans*. This tiny nematode has some of the same advantageous properties exhibited by the *A. californica* and *L. polyphemus* nervous systems. Most of all, compared with vertebrates, it is very simple. The number of neurons in the nervous system of *C. elegans* is also relatively small (302) and the axons interconnecting neurons unusually do not branch very much as they run between ganglia. As a result, only a few thousand synapses are known to interconnect the 302 neurons. Furthermore, not only are the same neurons present in one sample of *C. elegans* to the next, but also each neuron has been identified and both its shape and its location mapped. According to Sporns (2011), this is the only model creature for which such a complete nervous system map currently exists. The mapping was an enormous task carried out by White, Southgate, Thomson, and Brenner (1986) based on an exhaustive reconstruction of a series of electron micrographs. As a result of this complete mapping and some ingenious research techniques, *C. elegans* has proven to be a useful model animal for studying the relations between behavioral and neurophysiological processes.

However, this approach has not been without difficulties and challenges. Why should this be the case given the detailed nature of all of the neurons in the brain map we currently possess? The main answer to this question is that the *C. elegans* nervous system, despite its small size, is considerably more complex than the other two model nervous systems just discussed. The word "complexity," so easy to articulate and so difficult to explain, is an expression of a general principle permeating this entire book—the complexity and numerousness of the neuronal network of even these simple model animals pose combinatorial problems that quickly explode to demand unachievable computational powers.

The situation concerning the use of *C. elegans* as a model nervous system, therefore, is quite different than the ones characterizing *A. californica* or *L. polyphemus*, both of which have simplifying attributes. In the former, it was possible to distill the germane interactive network function down to a single repetitive functional process (mutual lateral inhibitory interaction) in which the arrangements of individual nodes, the ommatidia, were almost crystalline in their regularity. This made iterative computational algorithms simple enough to be executed. In *A. californica*, it turned out that the germane functional process (habituation) was actually ascribable to mechanisms that could be localized to a single interneuronal junction.

Furthermore, both *A. californica* and *L. polyphemus* are relatively large animals compared with *C. elegans*. Thus, they could be manipulated and probed with stimulating and recording instruments in a way that the smaller neurons of *C. elegans* could not. The ommatidia of *L. polyphemus* and the axons and ganglia of *A. californica* were large enough to be easily manipulated in an electrophysiological experiment. The small size of the animal made simple mechanical manipulation of *C. elegans* difficult—special techniques had to be developed to immobilize the creature's miniature worm-like body. *C. elegans* neurons were not only small but had the disconcerting property of bursting when punctured with a microelectrode. Nevertheless, the task of electrophysiological recording from this tiny creature was accomplished for neuromuscular junctions by Raizen and Avery (1994) and then for neurons by Lockery and Goodman (1998) and Goodman, Hall, Avery, and Lockery (1998). A useful compendium of practical advice detailing how neurophysiological recordings can be made from *C. elegans* can be found in an article by Richmond (2006).

Despite its small size, the resulting practical problems, and the relatively large number of neurons in this "simple" nervous system, *C. elegans* has been successfully used as a model for olfactory aversive learning. Ha et al. (2010) were able to compare changes in behavior as a result of experience with the operation of a relatively small neuronal network consisting of 13 neurons located in a ganglion in the anterior end of this nematode. The experimental procedure they used was based on an innate behavior exhibited by *C. elegans*; this tiny nematode responds to the activation of chemosensitive neurons in its head region to a large number of chemicals secreted by the bacteria that form its food (Bargmann, Hartwieg, and Horvitz, 1993). Interestingly, although many of these chemicals are detrimental to the animal, instigating diseases or metabolic failures, naïve animals that have not been exposed to these chemicals are attracted to rather than repelled by the chemical odorants. It is only after some experience with the negative effects of these chemicals that *C. elegans* avoids them. The task faced by Ha and his group was to determine how and where this model nervous system was organized to alter this behavior from attracted to aversive.

One of the first problems encountered was to measure the change in behavior from the naïve (attractive) to the mature (aversive) state. Fortuitously, the pattern of movement executed by *C. elegans* toward attractive chemicals differed from its pattern of movement away from aversive chemicals. There was a much smaller proportion of a kind of behavior Ha and his colleagues designated as Ω turns ("sharp turns in which the animal's body shape resembles the Greek letter omega Ω"), when the animal behaved as if there was an attractive odorant present, than when the behavior suggested that the chemical was aversive. Populations of *C. elegans* could then be trained to be used as controls for those that had been "taught" to avoid detrimental odorants and those that had not been. The proportion of Ω turns during attractive behavior could then be compared with the proportion during aversive behavior to determine whether the chemical was attracting or repelling *C. elegans*.

To explore the questions of the neural correlates of this learned behavior (from attraction to avoidance), it was then necessary to manipulate the *C. elegans* nervous system to determine the role of individual neurons in the learning process. This was accomplished by what was also an ingenious technique. Ha and his colleagues used a laser whose beam size had been demagnified to the point that it could target individual neurons. The purpose of the laser was to destroy targeted neurons. Therefore, any changes in *C. elegans*' behavior could be attributed to the absence of that particular neuron. The results of their experiments with this kind of selective destruction allowed them to map out the details of a network of 13 neurons that accounted for the naïve attractiveness of the odorants, the learned avoidance of them, and the experienced avoidance of the odorants. This analysis allowed Ha and his colleagues to uncover the fact that two different configurations of a relatively simple neuronal network

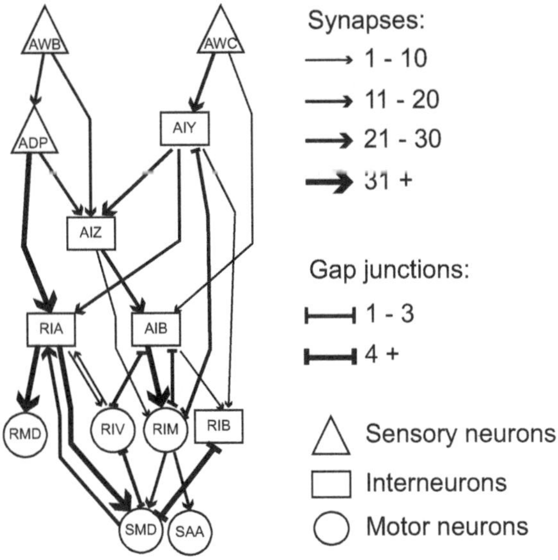

FIGURE 4.4 The neural network mediating the attractive–aversive behavior of *Caenorhabditis elegans* to odorants. Only 13 neurons control a complex behavioral pattern.

Reproduced from Ha et al. (2010) with permission.

mediated the naïve attractiveness and the trained aversion responses to the odorants. The network of neurons mediating the two forms of behavior as well as the learning process that changed the behavior is shown in Figure 4.4. Each of the blocks identifies one of the 13 neurons that are interconnected by the number of synapses indicated in the side bar.

Ha and his colleagues summarized the role of destroying various neurons on the naïve (attractive) and learned (aversive) chemosensitive behavior as well as the role individual neurons had on the learning process itself. With the pattern of observed neuronal interventions and behavioral changes they were able to construct the neuronal network shown in Figure 4.4—a remarkable tour de force made possible by the evolutionary accident of *C. elegans*' anatomy. The following list abstracts a few of the conclusions they drew from their experiments that allowed them to construct the neuronal network of *C. elegans*:

1. Learned aversive olfactory behavior requires the presence of the AWB and AWC olfactory neurons. The network shows one candidate network that regulates this naïve and learned behavior.

2. Two different neuronal networks regulate the attractive (naïve) and aversive behavior (learned), respectively.
3. The destruction of the AWB sensory neuron in one of these networks abolished the naïve attractiveness and learning.
4. The destruction of RIA interneuron in the other network abolished the learned preference and learning but did not affect the naïve attractiveness.

(Abstracted and paraphrased from Ha et al. 2010)

To cognitive neuroscientists, this work is of special interest because it is one of the relatively rare instances in which a model neuronal network has been directly manipulated and then changes in an animal's behavior measured. Of course, it cannot be a complete story since changing behavior from attraction to aversion is only one of the functions that must be carried out by even as simple a creature as *C. elegans* to guarantee the survival of its species. Other of the 320 neurons in its tiny body have to be involved in other behaviors. Furthermore, there are about 24 chemosensory neurons that are not involved in the particular behavior Ha and his colleagues have highlighted in their important and ingenious research. How far toward cognitive neuroscience we can search for a full decoding of even this intermediate example of a neuronal network is yet to be determined.

4.5 *Drosophila melanogaster* and Courtship Singing

No discussion of model organisms can be made without at least a brief mention of *Drosophila melanogaster*, the common fruit fly. This tiny creature was probably the first intentional example of the use of a simple organism to serve as a surrogate of more complicated creatures. Its first use as a model is attributed to Thomas Hunt Morgan (1866–1945) who showed that chromosomes were the vehicles for the transference of genetic information from generation to generation. He was able to accomplish this feat (for which he, too, won the Nobel Prize) because the genetic structure of *D. melanogaster* was much simpler than that of higher organisms—it consisted of only four chromosomes compared with the currently accepted human count of 46. Many of the genes on these four chromosomes have been associated with specific behavioral, chemical, and anatomical aspects of this simple creature. By manipulating the genotypes it has also been possible to associate genetic properties with observed behavior, thus making it relevant to cognitive neuroscience.

However, our interest is in neural theories and the literature describing the use of *D. melanogaster* in neuroscience is sparse. The reason for this lacunae is that the nervous system of this fly contains many more and much smaller neurons than are found in the *L. polyphemus* eye (~1000), *A. californica* (perhaps as many as 300 are involved in some way in the gill withdrawal response), or *C. elegans* (320). Estimates are that the *D. melanogaster* nervous system contains about 135,000 neurons compared with these other model animals. This number currently exceeds the threshold for the kind of neuron-by-neuron analysis that was possible in the other animals. Current efforts to map out the nervous system are limited to a few "compartments" and only 16,000 of the 135,000 neurons of the fruit fly's nervous system have been identified (Kohl and Jefferis, 2010; Chiang et al., 2011). To accomplish even this "coarse" and partial wiring diagram required that Chiang's group dissect more than a million fly brains! Clearly, this type of complete neuronal wiring diagram is probably not going to be possible for an organism as complicated as that of *D. melanogaster*, much less that of the human brain in the foreseeable future. Furthermore, even if we had a complete wiring diagram, the technical difficulties involved in repeatedly stimulating and recording from a specified neuron (or group of neurons) in the fly are formidable. The neuronal network of *D. melanogaster* clearly establishes a high limit for what we have designated as intermediate levels of complexity.

It is also important to remember that in the long history of using *D. melanogaster* as a means of studying genetic effects on anatomy and physiology, cognitive neuroscientists are doing something quite different when comparing behavior and genetic information. Olsen and Wilson (2008) clearly express this caveat in the following terms:

> Many classical Drosophila behavioral paradigms were designed to screen many flies simultaneously for profound [and easily observable] effects. But neuroscientists are increasingly interested in fly behavior for its own sake, rather than simply viewing it as a tool for isolating genetic mutations.
> *(p. 512)*

In other words, behavior is not nowadays being used as a tool for studying chromosomal effects; rather the goal is to explain the neural foundations of overt behavior. There is an important difference between studying the genotype and the neural or behavioral phenotype; the success that has been forthcoming in the genetic studies sometimes leads us to overstate our accomplishments in the neurophysiological domain. Furthermore, many of the properties that were advantageous in the other model preparations (e.g., small number of neurons, simple coding schemes, isolatable and identifiable neurons, or simplifying considerations like the regularity of the ommatidia anatomy in *L. polyphemus*) no longer apply.

Despite these disadvantages, some impressive neuroscientific work relevant to behavior has been done using *D. melanogaster*. The fruit fly has some advantages; for example, it has a wide variety of easily observed behaviors—eating, navigating, reproduction, etc. A major advantage (countering the increased number of neurons and anatomical uncertainties about its brain) is that the fruit fly has a much wider array of behaviors for which neurophysiological correlates can be sought than is available in some simpler animals. The price one pays for this expanded behavioral array is that the fly's nervous system is also more complicated and the difficulty in determining which neurons mediate which behavior has proven to be a serious constraint on the use of *D. melanogaster* as a model preparation in neural and behavioral comparisons.

The initial task, therefore, in planning an experiment to study a neural behavioral relationship is to determine which neurons of *D. melanogaster's* 135,000 neurons participate in a given behavior. Olsen and Wilson (2008) proposed three general methods to link particular neurons to particular networks:

1. Anatomical Method: Using a genetic manipulation to make a small number of neurons fluoresce when stimulated with light. The fluorescing neurons are assumed to be associated with an otherwise unknown set of interacting neurons.
2. Behavioral Method: Genetic manipulations may be used to selectively "silence" certain neuronal networks. If an associated behavior is "defective," the silenced neurons are assumed to be associated with the behavior. (Conversely, Flood, Gorczyca, White, Ito, and Yoshihara, 2013, have also developed methods for activating specific interacting neuronal networks. In this case, neuronal networks are activated rather than suppressed.)
3. Physiological Method: Presynaptic neurons are stimulated. The idea is that only synaptically connected neurons respond. The response in this case is activation of a calcium indicator (G-CaMP) in neurons that are functionally interconnected with the stimulated presynaptic neuron.

(Paraphrased from Olsen and Wilson 2008, pp. 513–515)

Using combinations of techniques such as these, it has been possible to make some limited progress in linking specific neurons with specific behavioral components in *D. melanogaster.* One of the most interesting experiments is a study of the control of the male courtship song of *D. melanogaster* carried out by von Philipsborn et al. (2011). Not surprising to any of us that have heard a fly buzzing around, this "song" is a well-developed behavior. It is, however, surprising that fly singing is controlled by a relatively small interconnected group of neurons that deals with different components of the song. One of the components necessary to evoke the male's song is a sensory signal from a female. In the absence of such a signal, it proved possible for von Philipsborn and her colleagues to induce a fly to sing by thermal stimulation.

The fly's song is a complex of a number of different motor responses. Two components of a "pulse song" include a sequence of from 2 to 50 pulses and a "sine song" consisting of a 140–170 HZ tone that together make up the neuronal foundation of a "decision" to sing (von Philipsborn et al., p. 509).

Five classes of neurons, all of which are genetically controlled by a single gene designated as fru^{m}, have been associated with the expression of the male phenotype of singing. Although these are not individually identified neurons (as many as 2,000 neurons may be involved in controlling this behavior), it has been possible to make some associations between groups of neurons and specific components of the song. The five types of neurons and the role each plays in the courtship song are discussed by von Philipsborn et al. and I briefly summarize the findings here.

Two of the five classes of neurons (P1 and plP10) are not involved directly in motor control of the singing behavior; instead, they seem to control the decision whether or not to sing. Neuron class P1, von Philipsborn and her colleagues asserted, was the key to triggering the pulse song—an analog of "decision making." It possessed a rather short axon and was limited anatomically to the region of the brain. Similarly, plP10 neurons were also involved in the decision to sing. The suggestion was that P1 was an input to plP10, conveying information from a distant ganglion since it (plP10) had a very long axon that terminated in a distant region of the brain.

The three other neuron classes that were involved in the fruit fly's courtship song—dPR1, vPR6, and vMS11—coexist in the thoracic ganglion of *D. melanogaster* and may, according to von Philipsborn and her colleagues, collectively encode the program controlling the specific properties of the song. Figure 4.5 is a theoretical diagram proposed by them for the connectivity pattern of the salient neurons in a network of neurons for the control of courtship singing. The array of challenges involved in carrying out an experiment such as this is impressive.

It is also important to appreciate that the theory shown in Figure 4.5 is of a much more conjectural nature than are the three models proposed in the earlier sections of this chapter.

4.6 Interim Summary

This chapter has reviewed an intermediate level of neuronal networks based on real neurophysiological data. Because of a variety of evolutionary accidents, it has proven possible to use the nervous systems of simple animals (usually invertebrates) as surrogates or models for more complex animals (usually vertebrates). Models, thus, serve as simple accessible idealizations of more complex systems. The individual utility of each model preparation is based on one simplifying condition or another that occurred in the course of organic evolution, the reasons for which we are only barely aware. Because of the relative simplicity of the observed neuronal networks, it has been possible to dissect (both functionally and anatomically) the simple neural systems of the model animals discussed here. It has also been possible to determine a modicum of the relationship between behavior and what are usually very simple neuronal networks. Indeed, all four of these applications of real biological model systems end up discussing networks that are almost trivial, thus suggesting that the direct study of real neurophysiological networks is not likely to be a promising strategy.

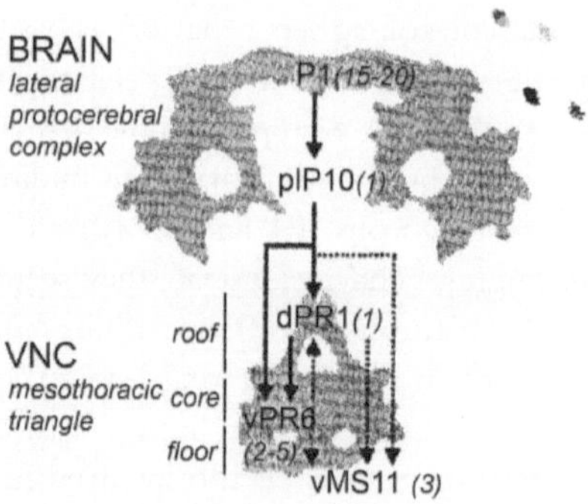

FIGURE 4.5 The neural network mediating singing behavior in *Drosophila melanogaster.* Five classes of neurons (not individual neurons) are involved. P1 and plP10 neurons control the decision to sing and dPR1, vPR6, and vMS11 control the temporal characteristics of the song.

Reproduced from von Philipsborn et al. (2011) with permission.

A prevailing assumption guiding this work is that these model systems are similar enough behaviorally to the behavior of higher-level organisms to at least justify drawing functional inferential analogies between the model and the modeled and, at best, assert that we have found the actual neuronal mechanism of what are common behaviors.

This line of research, however, is constrained by the fact that although the observable behaviors may seem to be the same, they may be actually instantiated by very different mechanisms in the model and modeled organisms. For all practical purposes these are bottom-up approaches to the role of networks and this type of research using conventional neurophysiological techniques fails for networks consisting of more than a few neurons.

Throughout this discussion, the limits of this direct neurophysiological approach have been apparent because the equivalence of the mechanisms underlying the model and the modeled behavior, respectively, remains uncertain. Although philosophers may blanch at the idea (referring to the logical fallacy of assuming something is true because it is possible), neurophysiologists will appreciate that there are innumerable, quite distinct mechanisms that could produce the same behavior. This harks back to the problem of underdetermination of the underlying mechanisms if we can observe only the behavior.

Among many other conceptual problems with this direct approach to studying neuronal networks, it appears that many investigators who used the model preparation approach seem to assume that not only do behavioral analogies exist between higher and lower animals, but also mechanistic identities in the form of neurophysiological homologies. Unfortunately, even the most robust data says nothing about the comparable neuronal network in the higher, modeled animal. That there are alternative possible and plausible neuronal network mechanisms that could produce responses in different animals based on totally different neuronal networks seems likely; indeed, our intuition tells us that there must be such alternative explanations. The advantage of this bottom-up approach is that it is anchored to neurobiological reality—the mechanisms described to explain behavior are based on real, tangible, and observable mechanisms.

An alternative and sometimes useful strategy to understand relatively simple intermediate neural networks is to work from the top down—to infer the possible nature of the neural network from an organism's behavior. Unfortunately, most such theories are functional inventions comparable to those operating at the psychological level. Models such as these are exemplified by the "biologically inspired cognitive architectures" (Goertzel, Lian, Arel, de Garis, and Chen, 2010, p. 30).

The important thing to remember about such top-down contributions is that they are not initially based on neural or neuronally based empirical data; they are functional mathematical descriptions in which the starting point is behavior and the end points are plausible mathematical and neural constructs—in other words, hypothetical constructs. The mathematical representation is invented to subsequently complete the description by showing a plausible manifestation of the behavior in neural terms—hypothetical neural mechanisms that could implement the functions described by the mathematics. These theories are not neurophysiological evidence-based but are top-down theoretical statements based on behavioral observations rather than neurophysiological ones.

A final and very important finding from the direct neuronal approach is that none of the methods or approaches discussed here was really successful at unraveling the details of a network. *Limulus polyphemus* permitted us to examine results distributed over a few receptors but that had to be simplified by the regular, almost crystalline, structure of the animal's eye. In the other experiments, it was not the network interaction between neurons that was studied but rather the role of specific and individually identified neurons. The "networks" studied were, at best, interactions between a few neurons. Interactions were just too complicated for our instruments, our analytical techniques, and our basic assumptions to tease out the details of neuron networks whose number exceeded the fingers on our hand. In short, only trivial "model" networks could be examined. Multiple electrode arrays could not reach our goal, nor would limiting ourselves to what were supposedly simplified (to the point of triviality) model preparations. Nor, for that matter would the construction of hypothetical entities inferred from top-down methodologies.

In conclusion, the use of model preparations with simple nervous systems has allowed us to unravel systems with a few neurons. However, this direct neurophysiological approach is severely limited to networks made up of very few neurons and cannot be expected to solve problems involving realistic networks.

What seems clear is that a direct neurophysiological approach to unraveling neuronal networks is not a promising approach. Although it is likely that we will learn much about neurons and synapses, neural networks will remain inscrutable from this point of view.

Looming on the horizon, however, was a new development that promised to break through the conceptual and technological limits I have discussed in this chapter. The development of the digital computer seemed to provide an ideal means of dealing with complex networks. The next chapter reviews the impact of these extraordinary devices on theory development in cognitive neurosciences.

Notes

1. It is likely that the phenomenon had been known years before. Edge enhancement occurs in the work of artists who antedated Mach by many years.
2. By symbolic, I am suggesting nothing more than that there might be neuronal mechanisms that are much more complex than the simple lateral inhibitory interactions accounting for the phenomenon in the human brain. We cannot exclude alternative explanations with the kind of experiments we are able to carry out in *L. polyphemus*—nor can we for any of the other model preparations discussed later in this chapter.

5

LARGE-SCALE COMPUTER SIMULATIONS (THEORIES) OF COGNITION

5.1 Introduction

In the previous chapter, I discussed an approach to studying neuronal networks that emphasized the examination of real, but small, neuronal networks using direct neurophysiological methods. The history of this bottom-up method makes one point very clearly: the complexities of even a very small network (consisting of from one to perhaps a dozen neurons) were such that any hope of understanding how a network worked with a conventional neurophysiological approach was not likely to be fulfilled with the tools that were available. It was only special cases (such as the Hartline Lab's work on *Limulus polyphemus*) in which certain simplifying constraints (repetitive regularity) could be invoked that permitted larger numbers of neurons to be considered. Even then, traditional neurophysiological methods do not seem promising—there are just too many neurons for standard analytical techniques to work.

Currently, however, new tools are becoming available that promise to change not only our experimental methodology but the concepts and assumptions that are certain to guide theory in particular in the years to come. Some of the most influential are certain to be modern versions of the digital computer. In principle and a priori, the tool "par excellence" for studying microneuronal networks is the digital computer, but not just the simpler machines of the class sitting on our desks. They may be useful for studying the intermediate level of neural networks of the kind described in Chapter 4. Instead, computers are needed that have the ability to manipulate huge numbers of computational elements, each one of which is capable of representing the activity of a single discrete element—an element such as a neuron—and to store astronomical amounts of data.

The potential solution to the challenges of numerousness and complexity in the microworld is a new kind of computer—the massively parallel system in which very large numbers of basic computing elements are physical or programmed mimics of neurons. The goal is to use a system such as a giant simulation device to study the collective information-processing properties of many neurons and to determine if such a simulation compares reasonably well with some aspect of cognitive behavior. Indeed, the microneuronal approach is often not even a mathematical or computational one—the "computer" is used as a simulator or model of a complex network instead of a formal mathematical engine. The details of this approach will become obvious as we review the relevant literature concerned with "supercomputer-based" theories and note the paucity of ordinary mathematical formulations used by their authors. Instead, the reports are filled with discussions of the properties of the networks that are constructed on them. The hope is that by understanding how simulations or emulations of large networks work, we may understand comparably sized networks of neurons.

Let's not get carried away, however, at this point. The goal of understanding the microneuronal network basis of the human mind is extremely ambitious; recall that the number of neurons in the brain is estimated to be about 86 billion. It would not be possible to pursue the goal of decoding the brain without the recent development of supercomputers that have the capacity to represent very large numbers of "neuromimes" and the operation of the even larger numbers of synapses estimated to participate in cognitive processes. Fortunately, as we see shortly, considerable progress has been made in engineering supercomputers of the kind that may be useful in Micro Connectionist Theory (μct) research. While it is certain that we will develop new computer technologies and programming techniques, it is no way nearly as certain that a compelling theory of mind–brain relations will emerge from the new kind of research that has been enabled by these primarily engineering achievements. As we see at the dawn of supercomputer research aimed at cognitive neuroscience goals, there is considerable controversy concerning the eventual outcome of this new approach.

For example, philosophers (e.g., Searle, 1997) express considerable doubt that a large-scale μct simulation, however behaviorally successful, would actually solve the most challenging problems of all—problems such as "can a simulation be conscious?" The problem is that the meaning of a "successful" simulation or explanatory theory of mind–brain relations is intertwined with some of the most important questions of human existence. It is for this reason that the goal of solving the mind–brain problem is of such widespread interest, even before we are able to determine what a "solution" would actually look like in a world in which we do not yet have a good definition of what mind or consciousness or cognition really are.

Thus, we have several goals in this chapter. First, I discuss the digital computer technology. I placed this brief introduction to this kind of technology at the beginning of this chapter because unlike some of the others discussed in previous chapters the digital computer plays a slightly different role than, for example, does microelectrode technology. Microelectrode technology provides a means of collecting the necessary data on which to build a bottom-up theory. The digital computer technology that is used in the theoretical applications to be discussed, however, is more like a top-down approach in which we seek to replicate a cognitive process with computer program and circuit tools. Understanding how modern supercomputers are used as theoretical engines in cognitive neuroscience is not directly tantamount to understanding many of the operations of the brain or solving the mind–brain problem. Indeed, as we see, most current μct supercomputer-based "theories" of the brain are actually descriptions of the computers on which they run rather than explicit theories of how the brain works. I begin, therefore, with a review of the technology of these extraordinary new supercomputers.

5.2 From Computer to Supercomputer

Serial Computers

The modern electronic digital computer represents one of the most influential technological developments in human history. Its importance ranks with the bow and arrow, animal domestication, the wheel, the microscope, and the automobile as vehicles for not only technological but also social change. Increasingly, in modern times, computers have affected the way we live and promise to have even a greater impact in the future. Especially in the past half-century, computer power seems to be evolving at a rate that could not have been imagined only a few years ago. Computer capability has grown so much that it now possible to dream, if not to achieve an actual simulation of human cognitive processes at the microneuronal level—thus activating a sea-change in cognitive neuroscience theory development. This is not being done at a symbolic or indirect level of analysis at which psychology and macroscopic theories operate, but at what many agree is the level of operation at which the brain actually accomplishes its wonderful, albeit mysterious, conversion of the neurophysiological to the mental. To understand the modern supercomputer fully, it is useful to give a brief history of some of the key points in its development.

Although computing engine-like devices of one kind or another have been used for many years (e.g., the Roman and Asian digital abacus, Napier's Bones, or the 17th century slide rule—an analog device) and a few mechanical calculating engines (e.g., the Jacquard loom) had been invented over the years, the modern electronic digital computer that will dominate the following discussion was first conceptualized and built by Atanasoff and Berry at Iowa State University—a usually overlooked history told by Burks and Burks (1989). The Atanasoff-Berry machine was an all-electronic, special-purpose, binary device capable of solving linear equations by numerical methods—the first of its kind. However, it had one critical missing feature: It could not be rapidly reprogrammed as it was hardwired to perform only a single function.

The critical idea—the storage of a sequence of instructions in the same memory as the data being processed so that both could be accessed and manipulated by the program—is attributed to both von Neumann (1945) and Turing (1950, 2004). Turing's design, although subsequent to von Neumann's, was considerably more detailed. Their joint breakthrough meant that a computer could rapidly (even during the course of a computation) alter its function. In this manner the computer became "general purpose" and metamorphosed into the universal information-processing machine that now underlies so much of modern science. The extraordinary idea underlying the great success of the digital computer is that decisions can be made by instructions that enable the programmer to choose the next instruction to be executed. Thus, what happens next could be contingent on what happened in the past. From this basic concept of contingent program executions ultimately came enormous powers of computation, simulation, and control. The von Neumann "computer," as it came to be known, thus became a general purpose device that belies its common name. As we see later, the supercomputer is only partially used as a computer; its role in cognitive neuroscience theory is overwhelmingly noncomputational, simulating neuronal networks and functions rather than arithmetical operations.

The von Neumann architecture, as it came to be known, has been the standard of computer design for many years. It was typically represented in the form shown in Figure 5.1.

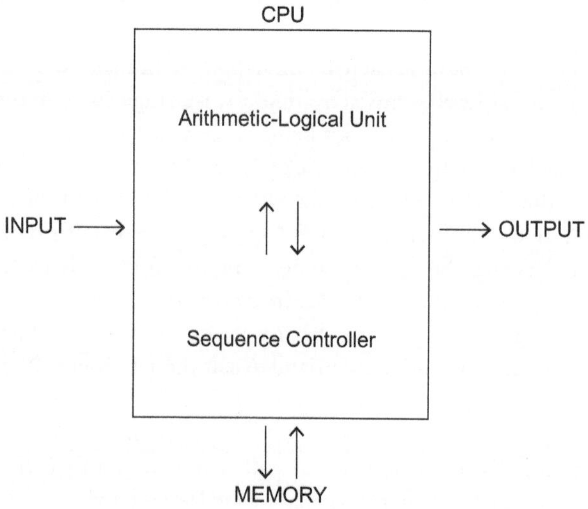

FIGURE 5.1 The organization of the single-core von Neumann computer composed of input device, output devices, memory, a central processing unit (CPU) that consists of an arithmetical-logical unit, and a program sequence step controller. Parallel processing computers may have many of each of these units connected in a variety of ways.

The block labeled INPUT consisted of any number of different devices that are able to take the information stored on some physical medium and enter them into the memory of the computer. Punched cards, punched paper tapes, teletypes, keyboards, cameras, and, nowadays, such devices as high-capacity "thumb" memories or highly sensitive cameras have been used as INPUT devices. Output devices such as mechanical printers or cathode ray tubes were among the early means of communicating the results of computations to the outside world. Nowadays such exotica as three-dimensional printers have begun to be used as output devices.

The central processing unit (CPU) was the computational device that accessed input data, manipulated it, and prepared it for presentation to the output devices. The CPU is often depicted as composed of two parts—an arithmetical-logical unit that actually executes the stored instructions and a control unit that tells the computer what instructions should occur next as it sequences the various steps of a program. Another name for the CPU was the "core," a term that rose in popularity as the older notion of a simple, single CPU began to be replaced by parallel processing systems that had multiple CPUs or specialized data manipulation devices that were quite different than the conventional single CPU of the classical von Neumann computer.

Memory technology evolved over the years from a novel use of Crookes tubes, to small magnetic spots on a drum, to vacuum tubes, to discrete solid-state circuits, to a huge variety of highly miniaturized and integrated electronic circuits, capable of rapid reading and writing of vast amounts of stored information. The development of fast printed memory circuits that could stably preserve information with or without power has been one of the grand successes of computer memory technology. Where originally a couple of thousand volatile memory locations constrained the size of programs, modern memories consist of gigabits of stable and rapidly accessible memory locations.

Early computers based on vacuum tube technology or early solid-state circuitry were ponderously slow by today's standards. They were all serial machines in which each instruction or computer access had to be completed before the next one could be executed. The main practical strategy to speeding up the execution of a computer program in the early years was to improve the speed at which memories could be accessed and information processed by the CPU. This led to an emphasis on engineering development of faster and faster technology (i.e., the electronic components); however, the basic organization plan of the von Neumann machine remained more or less constant, in general, if not in detail. Eventually, physical limits of heat generation and disposal and the basic physical limits on electrical conduction time (i.e., speed of light considerations) placed limits on how fast circuits in this kind of computer could function.

Refrigeration helped in part to overcome heat problems and new semiconductor materials helped to speed up computers, but eventually the physical limits encountered with very small circuits seemed to place limits on how fast a serial, single CPU computer could operate. Nevertheless, the first few generations of computers were all serial computers with a single CPU that was driven to the physical limits of available circuit design speeds. This did not mean serial computers were paralyzingly slow for their times; very large systems (e.g., the Cray-1 of 1976 with only a single CPU-dominated computer system speed until the 1970s) could accomplish much by concentrating on speeding up a single CPU.

We can sum up the extraordinary properties of digital computers by listing some of the most significant ideas that emerged during the past half-century that influenced the history, design, and applications of modern computer engineering:

1. Both program and data were stored in a common memory.
2. Random addressing of data and instructions was possible.
3. The program was accessible during its execution.
4. Programs and data could be modified during program execution.
5. This resulted in the ability to make decisions and modify the thread of the program sequence.
6. Both logical and arithmetical operations could be carried out.
7. Vacuum tubes and then solid electronic technologies permitted increasingly fast operation—but only up to a point determined by physical laws.

The physical limits on how fast a serial computer could process information were encountered just as the real world was demanding solutions to problems that required much faster computing speeds. Weather forecasting, molecular interactions, and a host of other civilian and military applications arose that could not be satisfied by the fastest available conventional single-CPU computer. The engineering solution that finessed the physical limit barrier was based on the concept of parallel processing—an idea that leapt into prominence in the 1980s. This is the topic of the next section.

Parallel Processing Computers

Many of the limits of serial, single-CPU computers were overcome to a considerable degree with the advent of multiple core or parallel processing computers. When it became obvious that the intrinsic speed of a circuit could not be pushed beyond fundamental physical limits, an appreciation emerged for how the effective speed of a computer with multiple CPUs operating in parallel might overcome these limits.

> A parallel processing computer is a device with multiple CPUs or cores so that multiple parts of a complex task can be carried out nearly simultaneously. Success with parallel processing depends on the task being partitionable into subcomponents.

The idea of parallel processing as a means of speeding up computer processing had been around almost as long as the idea of the programmable electronic digital computer. Surprisingly, one of the earliest computers, the Eniac (Goldstein and Goldstein, 1946), had some parallel processing capability in the form of several adders that could function simultaneously.

The idea of parallel processing, however promising it seemed to be, remained of secondary interest to computer engineers for many of the succeeding years, with the search for ever faster serial circuity dominating engineering thinking. The circuit design technology was advancing so rapidly that there was only an occasional need for seeking alternative approaches to improving computational speed. Even more inhibiting was a kind of psychological barrier among engineers—the notion that the ultimate speed limit was not going to be hit for many years. Increasing speed was associated with increasing circuit density and it had been predicted that circuit density was going to double every two years (Moore, 1965); empirically this seemed to be true. "Speed up," therefore, seemed unlimited into the foreseeable future as new materials and exotic physical principles were explored on which to base faster and faster circuitry. Occasionally (as noted by Gill, 1958, in one of the first articles specifically dedicated to a discussion of parallel computation), a primitive form of parallelization was built into a current computer design in the form of auxiliary devices that carried out some special computations at the same time as the single core was carrying out its computations. But, in the main, parallelization was not of high interest until the 1960s when several companies began to develop computers with a few duplicate cores or auxiliary arithmetical processing units. In addition, considerable concern was raised about the predicted difficulty in programming these multiple core systems. The goal of writing software to automatically partition a problem so that its parts are suitable for simultaneous execution and designing software specialized for parallel processing remains highly challenging.

By the end of the 1960s development of minimally parallel computers was off and running, with one system after another being offered commercially. It was in 1971 when the first dual-core processor was constructed on a single chip (by the Intel Corporation) that the massively parallel computer with many cores became technically feasible. Progress in printed circuit technology was rapid, but not instantaneous, and, as a result, even as late as 1984 the fastest computers (e.g., the Cray X-MP) still had only four cores. Whereas early supercomputers such as the Cray-1 sought to improve computing power by increasing the speed of the electronic circuitry of which they were constructed, the potential advantages of multiple core systems started to become evident. From that time till now, however, progress in parallelization has been extraordinary, with massively parallel computers now dominating the international speed race.

The first efforts to develop commercial multi-core and distributed memory parallel processing computers were relatively straightforward: two CPUs were mounted on a single "chip" and the computational workload distributed between the two. Special computational units in the earliest supercomputers were added to what were otherwise serial computers in an effort to speed up computational throughput. Later attempts to build ultrafast supercomputers made a dramatic change when it became possible to "print" large numbers of CPUs (the number now approaching millions of these no longer "central" but now better described as "distributed and parallel" processing system processing units). In recent years it has also been possible to build virtual supercomputers (e.g., the Beowulf system described in Becker, 1995) by taking advantage of world-wide communication links to employ distributed networks of ordinary desk computers to collectively solve the kind of problem that could be partitioned into separate tasks. The fastest supercomputers, however, still strive to package many memory and processing units in a single location; otherwise, communication times can become troublesome. The best solution to this difficulty is to concentrate on problems in which the partitioned components are totally independent. Unfortunately, nature does not often serve up its problems so neatly.

During the past decades of the 20th century, the idea of designing parallel processing computers with very large numbers of multiple cores and huge arrays of memory storage units became increasingly popular, and substantial investments were made by corporations and governments to engineer ultrafast computers. Parallel processing computers have evolved into a wide variety of system types; most fundamental differences are involved in the manner in which the multiple cores are interconnected with one another and with the memory storage units. Cores can be interconnected so that they are clustered together but still operate quasi-independently. Parallel processing computers also may have a variety of different means of accessing the memory. In the simplest cases, in which the memory is shared by the two CPUs, both might have full access to all of the memory, while in other designs only limited access to the common memory is available to each core. In a system in which two or more memories are connected separately to their respective CPUs, a common communication "bus" is necessary to let the two memories connect to each other. From a primitive two-processor, shared memory, the parallel processor has evolved a substantial variety of computer architectures in which multiple memories and CPUs are interconnected in complex ways by ever more complicated buses. Although it is impossible to provide a complete list of all possible parallel processing architectures, a simple and inclusive taxonomy was presented by Flynn (1972) that characterizes the main types. He categorized the main types but acknowledged that there were many variations on these main architectural types:

- Single Instruction Single Data (SISD)—The typical desktop sequential computer
- Single Instruction Multiple Data (SIMD)—One instruction affects all data
- Multiple Instruction Single Data (MISD)—Many instructions affect all data
- Multiple Instruction Multiple Data (MIMD)—The most common parallel processor

- Single Program Multiple Data (SPMD)—Multiple programs operating independently
- Multiple Program Multiple Data (MPMD)—Multiple programs under control of one core

(After Flynn 1972)

Speed is achieved in a parallel processing computer by adding and optimally organizing hardware—CPUs and memory—so that separable parts of appropriately parsed programs may be run as concurrently as possible rather than simply trying to speed up the operation of individual logical components. All other things being constant, a parallel processing computer should run faster on many types of problems than a single-CPU system as a straightforward function of the number of CPUs wired together in parallel.

However, there is a major problem built into the previous statement—all other things are not constant; in fact, parallelizing a computer virtually never speeds up program executions at the rate expected by simply counting the number of cores. The problem emerges from a limit on how the basic advantage of parallelicity can be exploited; namely the expectation that a program can be broken up into parts that can be run separately and the outcome patched together after the individual components (program parts) are all completed. The problem has nothing to do with the hardware or even the software of the parallel processing computer; the difficulty lies in the nature of the real-world problems being processed. Some tasks are intrinsically parallel: each step can be independent of previous or successive computations, that is, for some problems the order in which the various component parts are carried out does not matter.

An example of such an intrinsically parallel task is a computer vision system (such as a face recognition system) in which the various features can be processed independently and then combined to recognize a face. In tasks of this kind, each pixel (or larger area) can be examined by a parallel processing core independently to determine the respective parameters of a "face." The reason underlying the intrinsic parallelicity in this case is that the order in which each region is examined is inconsequential; each core in a parallel processor can be assigned to process a separate part of the image and the computational steps required for each part are independent of the others. When all parts of the image have been processed, the entire picture can be patched together, with only minor corrections of edge effects.

On the other hand, some problems are intrinsically serial and each successive part of the program depends on preceding ones. It is not possible to assign partial tasks to another CPU because the second part may depend on the evaluation of the first. Serial tasks have to wait until the necessary precursor program steps are completed before they can be executed. In short, each subsequent step depends on the result of preceding steps. An example of an intrinsically serial task is the computation of a Bayes theorem problem in which the prior values must be known before the post values can be evaluated. There is no way to jump ahead because subsequent steps are dependent on previous ones.

These idealized examples do not tell the full story because some tasks are only partially parallelizable. That is, in most problems, only a portion of the program can be broken down in a way that would permit additional CPUs to speed up the solution of the problem by bringing more raw computer power to the task. Indeed, there is a classical relationship known as Amdahl's (1967) law between the maximum possible speed increase and the proportion of the program that is parallelizable. A much simplified version of Amdhal's law asserts that the maximum possible speedup is less than 1/1-P for a given number of processors, where P is the proportion of the problem that can be parallelized. Thus, the maximum possible speedup of a task is not only a function of how many processors might be present in the computer (which in a perfectly parallel problem would be a linear function of the number of available cores), but also of the nature of the posed problem. As an example, if the problem was determined to be only about 50% parallelizable, then this law would predict that the fastest parallel computer would only be able to speed up program execution by a maximum factor of 2 over that achievable with a single ultrafast CPU computer. This would hold no matter how many CPUs were assigned to work on the problem.

In addition, the execution of a parallel program typically involves a number of additional housekeeping steps that were not required in a serial computer (e.g., assigning parts of the problem to particular subsets of the hardware and communicating data to multiple locations). In sum, one cannot expect to reach the theoretical maximum improvement suggested by Amdahl's law; there will also be some housekeeping costs to pay.

Nevertheless, with the extraordinary number of highly miniaturized cores that are now being combined into today's supercomputers, an enormous amount of computing power is becoming available that could not be achieved with simply technological (i.e., individual circuit speedup) changes. Some of the most powerful systems claim that they are composed of multiple millions of cores simulating tens of thousands of neurons. In addition to the economic, military, and scientific implications of high-speed computing, there are now quasi-political overtones and international competitions. Over the past few decades there has been an international race to build the "fastest" computer and for each competitor the strategy of choice has been to instantiate the contending computer in the form of a massively parallel system and then to test its performance with a standardized program, an example of which is Linpack, a program designed to solve a "dense system of linear equations."[1]

The metric of computer speed used to compare the contenders in this race is the unit referred to as Floating Point Operations per Second (FLOPS). As a raw measure of computer speed, the number of FLOPS is preferred to the number of machine language program instructions carried out per second because it represents a more realistic estimate of problem-solving ability, especially in floating point's ability to represent very small and very large numbers and fractional values (integer arithmetic deals only with whole numbers).

TABLE 5.1 Computer Performance Metrics

Name	*FLOPS*
YettaFLOPS	10^{24}
ZetaFLOPS	10^{21}
ExaFLOPS	10^{18}
PetaFLOPS	10^{15}
TeraFLOPS	10^{12}
GigaFLOPS	10^{9}
MegaFLOPS	10^{6}
KiloFLOPS	10^{3}

A typical desk computer today is capable of a few gigaFLOPS (or 10^9, or 10 billion floating point operations per second). Current state-of-the-art supercomputer speeds are astonishingly fast and getting faster yearly. The top reported speeds of the fastest supercomputers now available (e.g., the Chinese Tianhe-2) are reported to be in the petaFLOPS range (or about 10^{15}, or about a quadrillion FLOPS). A next goal of participants in the international competition is the construction of a computer capable of exaFLOPS, or 10^{18} floating point operations a second (a number for which we do not yet have a name, but which adds a certain informal substance to the word "zillion"). Table 5.1 summarizes the vocabulary used in quantifying computer speeds. The exponential growth in computer speeds in the past few years suggests that we might see a yetta-FLOPS computer in our lifetime, probably instantiated in some new hardware technology.

Whatever computer emerges that might serve as a test bed of an overarching theory of mind–brain relations, it is clear at this stage that even these high speeds will be stressed to their limits.

The Role of Parallel Processing in Cognitive Neuroscience

As modern neurology, neurophysiology, and many other theoretical and clinical investigators gathered more and more knowledge about brain activity, it became clear that mental activity is the result of brain activity or, in some more subtle manner, *is* brain activity. The correlations, if not the identity of brain activity and structure, on the one hand, and cognitive activity, on the other, have become increasingly obvious as decades of research and clinical experience provided a wealth of correlations between the two domains. Nevertheless, efforts to explain the specific relationship between brain and mind (in all of its ramifications) continue to fail for reasons that have occupied philosophers and theologians for millennia as well as psychologists and physiologists for centuries. The evolutionary emergence of human cognitive information processing, much less the inscrutable and inaccessible process we call consciousness or sentience, remain mysterious and covert.

Why should this be the case? One answer to this question lies in the complexity of both the brain and the mind. As noted in the preface to this book, somewhere in the neighborhood of 86 billion neurons and a thousand times as many synaptic interconnections make up the great neuronal networks of the brain. Today, most cognitive neuroscientists agree that it is in the actions of these networks—at the level of the neuron—that mind is encoded. That is, network operations executed at the neuronal level result in or become mental activity. However, the details of these network operations remain recalcitrant to theoretical explanation—there are just too many neurons and too few tools that seem to offer any hope of bridging the gap between the mental and the neuronal. As a result of this discrepancy between neuron numbers and analytical and technological tools, there was and continues to be a very sparse sample of satisfying mind–brain explanatory theories. Those that have been offered, such as the widely accepted Hebb conjecture, consist of some highly plausible conjectures that cannot be tested and remain speculative (with a few recent exceptions).

However, as just presented, the past few decades have seen an extraordinary technological development—massively parallel, multi-core supercomputers with enormous numbers of cores and memory storage units, which were described in the previous section. The parallel processing supercomputer represents the most recent technological development in the string of influential technologies that have successively guided cognitive neuroscience thinking. For the first time, however, there is the possibility that we might be able to study the operation of the brain or brain-like mechanisms at the microneuronal level of analysis at which cognitive processes are thought to be actually encoded. It is by no means certain that the progress we hope for will actually happen, but for the first time computer technology and brain science seem to be converging on a common, hitherto inaccessible level of analysis. For this moment in scientific history at least, there is a possibility of progress in mind–brain theory. Indeed it may not just be progress, but actual solutions.

However, this is still only a possibility and formidable obstacles lie between today's conjectures and the ultimate answer to how the brain produces mental activity. However, there are no guarantees that the "problem" will be "solved" by this new approach. It is not even clear what would be required to declare the "problem" solved. It is a new frontier of mind–brain science on which we have embarked in the past few years and where it will ultimately go is uncertain. Of one thing, we can be assured—this will be an exciting adventure.

The main reason that optimism is widespread these days is that there is a natural conceptual relationship between the modern supercomputer and the brain that did not exist with previous technologies. Both are made up of extraordinarily large numbers of basic units whose parallel processing is the main source of their joint power to manipulate information. Although the number of cores in even the largest parallel computer still does not come close to the 86 billion neurons of the human brain, it has been possible to use a smaller number to simulate numbers approaching that of the brain by means of iterative programming. Therefore, there is at least a superficial comparability that suggests that the basic functionality of the brain might scale down to that of the best modern computers or the computer might scale up to the level of complexity of the human brain. The current world speed holder—the Chinese Tianhe-2 supercomputer—consists of approximately 3 million separate cores of several different types. IBM's Sequoia supercomputer, the record holder in 2012, was built from 1.5 million cores.

It is important to distinguish among the number of electronic neurons, cores, and electronic circuits in a supercomputer simulation of a neuronal network. At the bottom of the hierarchy is the number of electronic circuits or gates. The number of these electronic circuits can be very large. Modern printed circuity has reduced the size of a binary flip-flop (the basic logical circuit) to molecular levels. Millions of these circuits are present in the most advanced supercomputers. On the other hand, a much smaller number of cores or CPUs are constructed from these basic circuits. Several million equivalents of cores seem to be the frontier currently. In turn, a number of cores must work together to represent a neuron so that a neuronal network can be simulated. About a million synthetic neurons is the best so far achieved. What is clear is that the number of actual physical neurons in the synthetic network is still significantly smaller than in the real biological brain and thus, repeated cycles of program execution would be required for the computer to run as fast as the brain.

Thus, there still remains a scale difference between the number of neurons in the brain and the number of cores in the most powerful supercomputers. It is not certain whether this difference will be significant or whether computers can be scaled up to the count of biological neurons. The important thing is that the past decade is the first period in cognitive neuroscience history that the basic organization properties of the technological tools available to study the brain and those of the brain itself are commensurable. The von Neumann serial computer under control of a programmable sequencer and a single computational element is fundamentally different in this regard from how the parallel processing computer and the brain operate. It was a poor model on which to base theory development, not only because of its serial nature, but also because of the difference in how the fundamental serial concept is embodied in the hardware. Although the degree of parallelicity is still far greater in the brain than the best computer, at least the basic principles of operation are now apparently converging. The basic concept defining the organization of a parallel processing computer and of the brain match much better now than with any previous technology.

Computers and the brain share another basic similarity that makes the parallel processing computer nearly the ideal tool for studying the brain. Unlike the serial computer in which each instruction is controlled in sequence by the commands of a "program," the parallel processing computer can, in principle, be far more autonomous. It is not necessary for a "program" of instructions to exist once the system has been set up; the momentary initial state of each of the simulated neurons in a parallel system of interacting cores can act as an initiator of activity. Thus, all the system needs to start running is an impetus—turning on the power or providing a metabolic source of energy might be all that is needed. Then, in accord with a set of rules of interneuronal interaction, it will continue to run out its normal course of action until it arrives at some final outcome determined by the interconnection pattern. The brain must be acting very much like this in basic principle; an initiating condition or state or stimulus triggers the action of a vast network of neurons that runs continuously until a behavior is selected and executed or the metabolic power source is discontinued. In neither case need there be an external program guiding the activity. Instead, there is just a sequence of complex parallel states that are subject to rules of interaction.

The main objection to the universality of the autonomous "free running" parallel processing computer is that the electronic parallel processing computer must be set up with the rules of interaction and the initial states of the neurons. However, this difference is secondary—once the states of the simulated neurons are established, the parallel processing computer and the brain operate in very much the same way, as a network of nodes interconnected by communication links whose activity is driven by the momentary state of the system rather than by a sequence of commands from some external source.

Another similarity between parallel processing supercomputers (and not the basic von Neumann computer) and the brain is that the information is processed in both systems by being passed simultaneously from one neuron, electronic or neurobiological, to another by multiple tracts. Feedback, feed forward, and lateral excitatory and inhibitory processing are all found in computer simulations as well as in natural brain organization. Regardless of the specific organization of a particular parallel processing computer, the modern supercomputer system is designed to achieve its speed or power by the simultaneous executions of many operations rather than by speeding up the action of one CPU.

On the dissimilarity side of this thesis, there is a vast difference between the instruction execution speed of a neuron and that of a typical solid-state core. Individual neurons are outrageously slow in the present context, operating on a time scale of milliseconds. Communication speeds along an axon are measured in tens of meters per second and there are additional delays inserted by synaptic connections. On the other hand, computers—even the most modest desk computer—operate at clock speeds (to choose one way of quantifying computer speed) measured in gigahertz (billions of cycles per second). Even so, the key to this kind of speed enhancement in the brain as well as in the most powerful supercomputers is massive parallel organization of multiple components.

In sum, parallel processing is a foundation organization principle of both modern computer engineering and how the brain has evolved. Parallelicity is the basic concept that permits the brain to carry out operations that would have been impossible with the intrinsic slow speed of the neuron. Single cell theories of cognition are, from this point of view, simply incorrect at the most basic conceptual level. It is not possible for any single neuron to carry out the information-processing speed represented by our amazing behavioral abilities without massive parallelicity. The hope is that the new technology will open the door to future understanding, perhaps even crack the mind–brain problem. However, there are formidable obstacles to the construction of an overarching theory of how the brain produces the mind.

One such obstacle is that even the most advanced supercomputers are currently limited in the number of individual CPUs that have been joined together; the 2 or 3 million that have so far been combined are still far short of the number of neurons in the brain. Complexity increases with numerousness, therefore, there is no assurance that the emergent properties we designate as cognitive will appear with the supercomputers that are available at the moment. It is likely, if not certain, that we may still not be at the appropriate level of analysis and may still have to depend on future supercomputers with many billions of synthetic neurons before we can claim a solution.

Another obstacle is the problem of interpreting the results—when will we know that we have built the desired overarching theory of mind–brain equivalence? The output of any system, brain or computer, is presented in some form of behavior—for example, a successful pattern recognition. It is possible, however, to produce behaviors by several different means (where the word "several" may indicate very large numbers; see the discussion of Hilgetag, O'Neill, and Young, 1996). We call these behaviors that are produced by different mechanisms and processes "analogs" of each other. The problem is that the externally observed behavior of a person or a computer is underdetermined; there are not enough data in the behavior to determine which of "several" possible mechanisms accounted for that behavior. Both computers and people can be directed to produce particular behaviors in innumerable ways.

How then do we determine that a particular underlying mechanism (a reductive theory) is actually explaining the behavior in humans and not just simulating? The answer to this rhetorical question is that there is no way of choosing among the analogs. However accurate a theory may be in producing a particular behavior, it may do so by mechanisms that are not only completely different than those used by the brain but may also be inaccessible to the most powerful of our tools. Recognizing a face, for example, can be accomplished by a computer by measuring the connecting lines between a set of key points (e.g., the tip of the nose) and then creating a numerical vector associated with that particular face. Humans also recognize faces but it seems extremely unlikely that we do so by comparable numerical calculations and measurements. Rather, cryptic (so far), organizational processes involving molar spatial relationships of the kind studied by the nearly forgotten Gestalt psychologists seem more likely mechanisms of face recognition. In principle, therefore, the actual neuronal underlying mechanisms of face recognition, like many other cognitive processes, may remain uncertain if all we have is behavioral data and knowledge of how large-scale parallel processing computers work.

Again, it cannot be stated too often that the availability of a new tool such as parallel processing supercomputers does not guarantee success in the search for understanding; however, it does offer an attractive promise that was not available until now. How far we can go in solving the mind-body problem using this powerful new tool remains to be seen.

Now let us briefly consider some of the parallel processing theories of cognitive processes.

5.3 Parallel Processing Theories of Cognitive Processes

Some History and Background

The foundation concept that the cognitive functions of the brain are embodied in a parallel processing network predated the availability or understanding of parallel processing computers, and indeed of computers in general. What was probably the first explicit model of a parallel processing network in the nervous system was the work of Pitts and McCulloch (1947) reported in their classical article on "universals." Among the earliest reports in which parallel processing networks were studied neurophysiologically was that by Hartline, Wagner, and Ratliff (1956) concerning the fortuitously simple organization of the horseshoe crab eye. (See the discussion of this research in Chapter 4.) As we saw there, because of the large size of the receptor cells in this compound eye, they were able to neurophysiologically examine the lateral inhibitory interactions between the neurons located in the horseshoe crab's ommatidia and explicitly evaluate a real neurophysiological parallel processing system. Hartline and his colleagues demonstrated that this primitive form of parallel processing could neurophysiologically mimic the human perceptual phenomena of Mach Bands—the enhanced edges that appeared despite the absence of any corresponding luminance differences in the stimulus. Although a few computer scientists who had access to computers subsequently modeled this process, to the best of my knowledge, no reports of a computer simulation of the horseshoe crab eye were published during this early period. Nevertheless, the fundamental concept of parallel processing was established by these pioneering studies; parallel processing became a part of the theoretical culture that followed.

The idea of parallel computations as the basis for cognitive information processes such as learning took a major leap forward with the work of Rosenblatt (1958). He invented a device designated as a "perceptron" that was essentially a parallel processing system consisting of a small number of elements that could learn to discriminate between different stimulus patterns. Rosenblatt's theory set off a tsunami of activity in which parallel processing, perceptron-like devices were studied by a large number of theoreticians, including Selfridge (1958), Kohonen (1977), Hopfield (1982), and Hinton and Sejnowski (1986). Unfortunately, this work was largely, albeit temporarily, suspended when it was shown by Minsky and Papert (1969) that simple perceptron-like networks were intrinsically incapable of certain problems that were basic to machine learning.[2] The study of these simple networks was rejuvenated when Rumelhart, Hinton, and Williams (1986) defined the back propagation technique—an iterative parallel processing procedure in which parameters were constantly adjusted, thus overcoming some of the limitations raised by Minsky and Papert (1969). Most of the perceptron-like devices consisted of simulations of less than 100 neurons, beyond which many of the systems seemed to saturate, that is, all of the neurons became active. Study of these toy systems today is essentially the study of parallel interactions within networks composed of modest numbers of linear components, but has drifted far from its neurophysiological roots.

One branch in the historical development of the parallel network approach to theories of cognitive processes emphasized a network of functional cognitive units that were quite unlike the neurophysiological ideas that characterized some earlier models. The nodes in the networks proposed by the PDP Research Group (McClelland, Rumelhart, and the PDP Research Group, 1988; Rumelhart, McClelland, and the PDP Research Group, 1988), for example, were macroscopic behavioral units rather than either macroneural or microneuronal components. This approach came to be called "connectionism" and is still a powerful force in cognitive theory. Although it emphasizes parallelism, connectionism is almost totally free of neurophysiological concepts and concentrates on symbolic, functional, or behavioral nodes rather than explicit macroneural or microneuronal networks. Progress in this line of theory has been reviewed by Goertzel et al. (2010) and is discussed in detail in my previous books (Uttal, 2016).

Although the concept of parallel processing expanded to permeate the most advanced neuroreductionist theories of cognition, there was a major missing factor inhibiting progress in theory development at the end of the 1980s. Most of the theories and prototheories that I have discussed so far and will discuss in this section had to be executed on modest-sized serial computers—parallel processing supercomputers did not yet exist or were not available to researchers interested in modeling cognitive processes. For those few models developed and carried out on serial computers, parallelism was simulated by reiterative methods in which each component in a network was evaluated in serial order while the processing of every other component simply stopped and awaited its turn for attention by the single CPU. Obviously, however, this kind of iterative seriality is a time-consuming process even if the fastest available high-speed computer was used. The result was that whatever models were tested had a relatively small number (<100) of components in the networks they simulated.

As the end of the 20th century approached and passed, this situation began to rapidly change. Massively parallel supercomputers became available, and scientists and engineers of all persuasions quickly took advantage of their presence to execute large-scale simulations.

As promising as the parallel processing supercomputer is, in principle, as an engine of theory building, it is also possible to foresee some of the difficulties that are being faced and will continue to be faced in their application to the task of building cognitive neuroscience theories. Some of these difficulties are subtle and some are naïve confusions between the search for an explanatory theory, on the one hand, and the engineering of bigger, faster computers, on the other. Merely noting that a computer engineering project is "inspired by" a brain property does not a mind–brain theory make.

An initial problem facing theoreticians who use massively parallel processing computers as their tool of choice to theorize about cognitive or brain activity is that it is difficult to know what would constitute an acceptable explanatory theory. There are two source of this difficulty. First is the fact that both the overall internal states of a computer and mental activity are largely inaccessible. We have only the most indirect indications—mainly interpersonally observable behavior or introspective reports of the cognitive concomitants of a brain activity. However, neither of these measures permits access to the lower-level mechanisms that constitute the "meat and potatoes" of any reductive theory.

Thus, the most abstruse entity of all—consciousness—remains poorly defined and immeasurable as well as inaccessible; neither overt behavior nor covert thought provide any direct indication of what are the underlying mechanisms, either neurophysiological or hypothetical cognitive constructs that correspond to our awareness. As noted repeatedly in the course of my writing, behavior is underdetermined—it can be produced by what may be an innumerable number of possible internal mechanisms and there is insufficient information in a behavioral observation to determine the actual nature of those mechanisms. Therefore, we may arrive at a point where we have a *sufficient* answer to the basic query but no way of providing the *necessary* answer. If *plausible sufficiency* is acceptable to some theoreticians as the criterion for acceptance of *a* (or *the*) theoretical explanation, then so be it. It must be remembered, however, that *plausible sufficiency* is not too distant from the pejorative term "folk psychological theory"—it is just one of a possible multitude of mechanisms that can produce analogous behavior. One implication of such an interpretation is that if some kind of "necessity" or "truth" is desired, our search for a theory may be a long one.

Another major difficulty in basing theories on computers is that the very processes in which we are interested may emerge at a level of system complexity that exceeds even the most powerful computer's ability to manipulate the relevant information. That is, whatever the cognitive process under study, it simply may not be present at the levels of analysis at which we are capable of problem solving. To make this subtle point more clearly, whatever consciousness is, it may not exist in a network (neural or otherwise) until a sufficiently high level of network complexity exists and then, for computational reasons, it may be impossible to either produce or evaluate an overarching theory of the mind. Only when network complexity rises to an unknown, but extreme level, may a cognitive process actually become manifest. One has only to compute the factorial of a relatively modest number to appreciate that technological and combinatorial inadequacy in the face of realistic complexity may persist as a problem for cognitive neuroscience into the foreseeable future. (For a more complete discussion of this issue, I refer my readers to Chapter 3.)

I am most interested in the critique of the strategy of trying to draw overall conclusions of the individual's nature from studying the social algorithms.

> . . . looking at the algorithms [components] will not yield much insight, because the interplay of social and behaviors yields patterns that are fundamentally emergent. These patterns cannot be gleaned from reading code.
>
> *(Lazer, 2015, p. 1090)*

Although extracted from a completely different field of investigation, the point is that it may be infeasible to use the components of a system to predict its overall behavior because the characterization one is looking for only occurs when all of the pieces have been combined into the whole system. Although the "sum of the parts" may, in principle, be no greater than the whole, the whole may not become manifest until the parts have been integrated into a whole. Thus, the idea of trying to produce an overarching theory or explanation of something like the mind by emphasizing the nature of the parts may be a false hope. The mind (or whatever else one wishes to call it) may simply not exist in the context of its separate parts but only when all of the parts have been accounted for. This paradox may make any effort to analyze the mind unachievable.

This discussion clarifies one point. No matter how powerful a parallel processing supercomputer may be nor how plausible the parallel processing approach may seem, it is not certain at this point that raw brute force computing power will be the research vehicle that ultimately solves the mind–brain problem. Nor, for that matter, can we yet say for certain that the mind–brain problem is solvable. The residual question is—how would we know a complete, overarching, neuroreductionist theory of mind should we ever encounter it? Simply simulating a behavior will not produce a convincing theory because of the multiple ways in which a behavioral response might be implemented neurophysiologically. Presumably, such a theory would at the least require that we were able to determine the exact state of an ensemble of neurons and to demonstrate the necessity of that state for producing a behavior. It would also require that we are able to produce changes in the behavior by manipulating that state of the ensemble of individual neurons.

The question of the criteria for acceptance of a computer simulation as a theory of mind remains salient. Martin, Grimwood, and Morris (2000), in addition to their list of experimental *strategies* that would be necessary to support a neuronal theory of learning (which I paraphrased in Chapter 3), also provided us with a list of the empirical *outcomes* that should be expected if we are to support the basic hypothesis that patterns of network activity are the psychoneural equivalents of cognitive processes. Once again, I paraphrase their very important suggestions about the possible general findings (originally expressed by them in terms of learning) in a more general way to emphasize their relevance to other cognitive topics:

1. Correlation: Activity in neuronal networks must correlate with specific cognitive processes in at least some ways.
2. Induction: Changes in cognitive processes must produce changes in neuronal activity and vice versa.
3. Occlusion: Saturation of neuronal network activity should block cognitive processes and prevent learning new ones.
4. Intervention: Blocking relevant neuronal activity by experimental means should block related cognitive activity.
5. Erasure: Erasing a particular pattern in a neuronal network should eliminate the associated cognitive process.

(Paraphrased from Martin, Grimwood, and Morris, 2000, p. 652)

The point of this list in the present context is that not all of these expected outcomes may be available to researchers for both practical and conceptual reasons. Nevertheless, it is at this level (that of the individual states of the underlying individual neurons and synapses) that the essence of our cognitive activity is most likely to reside. It is not clear, for example, how one could actually carry out an experiment in which the experimenter had sufficient control over the independent variables to erase or produce a particular pattern. Experiments calling for "induction" or "saturation" sound very much as if they were macroneural manipulations rather than the microneuronal network manipulations that characterize parallel processing networks of the kind with which we are concerned in this chapter.

There is another very important question that has to be considered about each parallel processing theory—what are the respective goals of the various supercomputer "brain" models considered here? Although this book is aimed at considering those that specifically link to cognitive and neurophysiological studies, this is not necessarily the basic motivation of all of the models discussed here. Some, as we shall see, are primarily interested in large-scale simulations of neuronal networks, some with the production of particular useful behaviors, and some in the engineering of massively parallel computer systems themselves. The varied motivations cover a wide range of network interactions. Sometimes, the links are limited to just borrowing a few words from psychology or the neurosciences. Thus, computers were early on referred to as electronic "brains" despite a complete absence of any neural ideas beyond the fact that computers and brains were both "information-processing" machines. This was sufficient to create a metaphorical relation between the two ("computers are brains"), but insufficient to establish that an explanatory model of the transformation from brain processes to cognitive processes had been produced. Later, as it became obvious that the brain was built on quite different principles than the von Neumann computer, the serial computer metaphor slowly fell out of flavor.

Computer engineers, in their never-ending search for more powerful and faster computers, often looked to a popular computer metaphor for the brain—the parallel processor—for insights and suggestions about the organization of such systems. To these computer engineers, their efforts were "inspired" by the brain, but they were not intended to be explicit simulations of it. In some cases, this allusion to the brain often went no further than an acknowledgement of its most superficial properties (e.g., parallelism itself).

The point is that engineering parallel processing computers is a complex task in its own right, with its own problems and goals, many of which are irrelevant to the cognitive neuroscience task. It may be reasonably argued that the relation between parallel processing computers and the nervous system is still metaphorical rather than based on any robust empirical or conceptual links. Clearly, the materials from which the two devices are constructed are different and these material differences may produce different information-processing strategies. According to Hertz, Krogh, and Palmer (1991) the engineering problems associated with computer parallelicity can be summed up by the following questions:

1. What is the best architecture?
2. How can a network be programmed?
3. What can the various types of networks do?
4. How can a network be built in hardware?

(From Hertz, Krogh, and Palmer, 1991, pp. 8–9)

None of which has anything necessarily to do with the brain or cognition beyond the initial metaphor of their joint parallelicity. The overarching goal in this context is to build the fastest, and thus the most powerful general-purpose computer system—regardless of how the brain has evolved. Beyond the patently computer engineering projects, there are also a number of studies using parallel computer systems that are aimed at a neuron-by-neuron representation of known neuroanatomical structures. In this category, the goal is to observe the structure and neurophysiological responses of a network that exhibits comparable neurophysiological behavior to a structure such as a "neocortical column"—a basic anatomical unit now known to characterize the cerebral cortex. Others have had as their goal simulation of the interactions that may occur between other anatomical structures such as the thalamus and the cortex. So far, however, in this listing, explaining cognitive processes has been almost completely absent.

Another class of parallel processing computer research, however, is aimed specifically at comparisons between computer architecture and cognitive processes. Nevertheless, in this class of research, there is less interest in neurobiological fidelity than in producing comparable behavioral outcomes. In this context, the goal is to make computers behave like people, but not to restrict or constrain programs or hardware algorithms to the same computational processes used in the brain. Needless to say, this imitation, simulation, or emulation (and often enhancement) of human behavioral processes has been the main thrust of modern applications—to supplement or replace many of the tasks that previously had been carried out exclusively by humans without necessarily reproducing the same logical "programs."

As one considers this spectrum of motivations and despite the enormous interest in linking brain anatomy and neurophysiology to cognitive processes, it is important to appreciate that no *cognitive neuroscience* problem of mind and brain relationship has been resolved so far by the expensive and complex research that has been forthcoming; most theories of computer organization that have any relevance to cognition in computer science are still "toy problems." Clearly, there is a multitude of ways that computers can simulate human behavior that do not require the construction of exactly the same computer operations and procedures that the brain presumably uses. Simulations and imitations, no matter how perfect, cannot by themselves provide a necessary explanation of how the brain works to produce cognitive activity—merely a sufficient imitation of behavior. To the contrary, the ultimate theoretical goal of cognitive neurosciences is to understand the brain well enough to mimic behavior to a degree of detail that permits us to assert that we have solved the mind–brain problem.

This leads to another strategy in modeling cognitive processes with large-scale supercomputers—what we may refer to as the "hopeful" approach. The technique embodied in the hopeful strategy is to develop a simulation of a prototypical neuron and then connect large numbers of them together and just see what happens. It is probably the case that more of the parallel processing prototheories currently being considered fit into this category than is admitted by investigators in this field. Why this should be the case is obvious—although we know a lot about neurons and synapses, we know very little of the rules governing their transition to behavior. On the positive side of this "hopeful" scenario is the fact that an enormous amount has been learned about the engineering organization and performance of large-scale, parallel processing supercomputers.

There can be no dispute, however, that the most likely tool currently available to achieve any of these goals is likely to be some future descendent of the parallel processing supercomputer we know today. In this cognitive neuroscience context, however, the primary goal is to solve the problem of mind–brain correspondence, not to engineer the hardware itself. The use of supercomputers to achieve this goal is the current frontier. Projects are just starting that have the specific goal of providing an answer to the ancient conundrum—how does the brain make the mind? Only time and extensive research will determine if this is a qualitative paradigm shift or just another false alarm.

Obviously there is an increase in conceptual difficulty as the goals of research become more relevant to cognitive neuroscience—particularly, in light of the stated difficulty of determining what constitutes an acceptable theory of mind–brain interaction. Indeed, it is often not clear where each of these alternative strategies begins and another ends. It is equally possible that some engineering development will inadvertently "crack the code" for cognitive neuroscience as an "unintended consequence" of simply being active in the field as it is that some neurobiological discovery will suggest a new way of engineering higher-speed computers. Nevertheless, whatever the nature of the symbiosis between engineering and neuroscience, it must be appreciated that not all computer engineering projects, no matter how inspired by the brain, are intended to be theories of the mind–brain relation.

In sum, it is important to remind my readers once again that not all parallel processing computer projects to be discussed here are relevant to theories of the mind–brain relationship motivating this book. Simply using neurobiological or cognitive vocabularies does not automatically make either the hardware or the software into a theory nor does claiming that a parallel processing computer was "inspired" by the brain. There is, however, an optimistic note. Regardless of the goals of one of these parallel processing research projects, whether neurobiological or engineering, there is a common thread of activities, actions, and accomplishments. Each of the projects now to be discussed are based on a day-to-day technological foundation of parallel processing and each, to various degrees, does attempt to link their work to either neurophysiological or psychological parameters, albeit in what may be only the use of a few words.

Parallel Processing Computer Theories of the Brain

In the following sections of this chapter, I review some of the most significant of the recently reported large-scale computer applications to brain science. The criterion that has guided my selection of these exemplar studies is that they are based, in principle, on two fundamental assumptions—namely, massive parallelicity at the microscopic level and the idea that this mode of organization is common to the brain and the supercomputer. As we progress through this mini-review, it will become clear what some of the advantages and disadvantages are of this bifurcated approach to theory building in cognitive neuroscience.

An Early Model of the Thalamocortical System

From the point of view of neuroscience (as opposed to that of large-scale computer engineering), the neural network simulation project carried out by Lumer, Edelman, and Tononi (1997a, 1997b) was one of the earliest to attempt the challenging task of simulating neural networks on a digital computer. The immediate goal of these investigators was to develop a network model of the thalamocortical system—a group of brain regions previously having been shown to be involved in visual perception. Their long-term goal was to study how manipulation of the properties of a simulated neuronal network affected the behavior of the network and thus might provide some insights into the perceptual processes that were assumed to be carried out by neuronal interactions in this portion of the brain.

Specifically, Lumer, Edelman, and Tononi observed that there were two regions of the visual cortex (a primary visual area and a secondary region, each of which was organized into a three-layer column) and corresponding primary and secondary regions of the thalamus involved in their simulation. The cortex was subdivided into three horizontal neuronal levels and the thalamus into two layers in accord with the known anatomy of these regions. Both inhibitory and excitatory connections between the regions were simulated as well as both ascending (centripetal) and feedback (centrifugal) connections.

The Lumer, Edelman, and Tononi simulation fleshed out the detailed microneuronal structure of the various regions by simulating the activity of 65,000 neurons and 5 million synaptic connections. The model was run on a "general-purpose object-oriented neural simulator" (1997a, p. 211), programmed on an unpublished software system, and was executed on an otherwise unidentified, but, given its date, presumably serial computer. The primary method was to create an initial state of the network by stimulating it with grating patterns to initiate the network activity and then, by pooling the responses of the involved neurons, describe system-wide activity.

Lumer, Edelman, and Tononi's main goal was to examine the effect of manipulating various parameters of the network on the generation of network-wide oscillations. The parameters (which serve as the independent variables of their computer experiments) included such factors as synaptic strength, transmission delays, inhibitory time constants, and the structure of the connecting loops between the 10 regions (six cortical and four thalamic) being simulated. To observe the cumulative response of the system they averaged the individual neuronal activities at each of the regions of their model, seeking information about how each of these parameters affected the oscillatory behavior. Lumer, Edelman, and Tononi also carried out experiments in which "lesions" were made by removing the connections between the simulated brain regions represented in their model. The results of their experiments identified important features at both the individual cellular and "macrocircuit" levels that determined the kinds of oscillatory behavior resulting from these interventions.

Lumer, Edelman, and Tononi's report is especially interesting because it is unusually complete in spelling out the details of their goals, methods, results, and conclusions. It is clear how they go about defining the initial conditions—the stimuli—and the cumulative or population-averaged response measure they used to examine a very specific property of networks—its oscillatory behavior. From the neuroscience point of view it is anchored to well-established neurophysiological data concerning the structure and function of the early portions of the visual system. Thus, it added a significant amount to our understanding of how neural networks might be affected by measurable properties of the nervous system. The anatomical properties of the model correspond to well-known properties of the visual system such as layering and feedback.

On the other hand, their model is limited, as they themselves point out, in that it does not link directly to many of the problems of perception—instead it is a test bed to study biologically plausible networks of a well-defined portion of the nervous system. But they have not correlated these plausible visual neurophysiological properties with psychophysical data. In their words:

> While the simulations presented in this paper do not yet address functional questions . . . they provide a powerful tool to *begin* investigating these [perceptual] phenomena in closer contact with experimental observations.
>
> *(Lumer, Edelman, and Tononi, 1997a, p. 224, italics added)*

From a computer science point of view, despite what must be considered by today's standards to be a relatively small simulation in terms of the number of simulated neurons and synaptic connections, this pioneering work was extremely fruitful in establishing a methodology that guided later investigators to study and manipulate simulated neuronal networks.

In sum, the Lumer, Edelman, and Tononi neuronal network model was a project aimed at simulation of the neurophysiological foundations of a perceptual mechanism by a system of 65,000 computer-generated neurons.

Spiking Neurons

Another early project in which the stage was set for the more recent and much larger computer simulations was reported by Izhikevich (2003). Izhikevich programmed a relatively simple network model in which each simulated neuron (of a set variously mentioned as including 1,000, 10, 000, and "tens of thousands of simulations") was represented by a set of simple differential equations. The simulated neurons were interconnected by up to a million simulated synaptic connections, 80% of which were excitatory and 20% of which were inhibitory.

Izhikevich's goal was to simulate the spiking behavior of a slab of neurons comparable to the number in the rat's motor cortex. This was one of the first models that attempted to start at the level of the Hodgkin-Huxley membrane equations. The measure of neuron activity was an integrated value plotted on a time-spike occurrence chart from which the pattern of spiking activity could be used as a measure of the frequency components of the EEG (e.g., alpha and gamma rhythms) as a response measure. The activity in this slab of simulated interconnected neurons was triggered by initiating activity in a small set of the excitatory synapses.

Although all of the neurons were identical, the pattern of synaptic interconnections was random. The fact that there was no "learning" or "adaptive" responses of the synapses programmed into the model meant that the emerging waves of activity in the network led to orderly patterns of dynamic neuronal activity in the simulated tissue. Izhikevich argued that his simple model reproduced complex wave-like activity typical of the organic brain that included "spiking, bursting, and mixed modes" (p. 1572). Another interesting aspect of his model was his claim that, because of the relatively simple properties of the simulation (the structure of the model's neurons was minimally represented, each by a common differential equation), a system of "tens of thousands" of neurons and a relatively large number of synapses could be run on a small PC-class computer, albeit presumably not in real time. Nevertheless, the activity of a simulated slab of brain tissue was complex enough, even with the constraints of the simplified parameters of the model, to reproduce complex wave-like patterns of coordinated brain activity.

Izhikevich's pioneering work set the stage for some of the simulations that were to follow based on his idea that patterns of spiking activity would spontaneously arise as a function of the initial arrangement of the synaptic junctions. Subsequently, the Izhikevich model was expanded by Izhikevich and Edelman (2008) to a much larger-scale simulation of the thalamocortical system, with a greatly enlarged and detailed simulation of 22 different kinds of neurons, as well as more specific connections among a larger number of simulated brain and brain stem regions. A number of other properties of the network were discovered, not the least of which was its sensitivity to the response of a single neuron and its tendency to turn off rather than to saturate. (This latter property was probably a result of the proportion of excitatory synaptic junctions.) Izhikevich and Edelman also developed novel display techniques that helped investigators represent the overall picture of the network's simulated neuronal responses.

In sum, the Izhikevich and Izhikevich and Edelman simulations involved "a million neurons, tens of millions of neuronal compartments, and nearly half a billion synapses" (p. 17) whose goal was to simulate the activity of a large neuronal network. The model did not run in real time even on the 60 core parallel processor available to them.

The Neurogrid Project

Another example of a network research project in which a large number of parallel organized neurons was simulated was the Neurogrid project (Boahen, 2006; Benjamin et al., 2014). Avowedly drawing inspiration from brain circuitry, this group used a combined analog and digital circuit technology with the goal of building a power-efficient, parallel processing computer that simulated the activity of a million neurons and 6 billion synaptic junctions. The Neurogrid project was unusual in that it attempted to carry out the simulation in real time, an advantage then made possible by the partial analog technology used by its originators.

The end product of this engineering development project was intended to be a computer hardware system that could be easily programmed without any specific knowledge of the networks that instantiated the hardware. The intended applications of this hardware were primarily in specific engineering applications to robotics and sensory prostheses. Although paying some initial lip service to the goal "understand[ing] how cognition arises from neuronal properties" (Boahen, 2006, p. 9), all of the attention of this project was directed at engineering application. Nevertheless, as I noted earlier, it is sometimes difficult to distinguish between engineering and cognitive neuroscience applications when one examines research in parallel processing networks.

In sum, the Neurogrid project was a hardware development endeavor aimed at producing an easily programmed, power-efficient, parallel processing system consisting of a million simulated neurons that involved both analog and digital circuitry.

The Spaun Project

An ambitious attempt to specifically solve some of the most challenging of cognitive neuroscience problems by building a parallel processing computer model of a cognitively competent system has been presented by Eliasmith et al. (2012). This group's goals were to show how a large-scale simulation of a brain system could link to behavior—a connection that these investigators believed had generally not been accomplished in most of the earlier work. In their words, the goal was "to relate the incredibly complex behavior of animals to the equally complex activity of their brains" (p. 1202).

To achieve this goal, Eliasmith and his group programmed a software system called Nengo to simulate 2.5 million neurons to carry out one of eight kinds of cognitive tasks, examples of which were "copy" "to recognize" an input alphabetical character. Stimuli were either handwritten or typed characters and responses were "the movements of a physically modeled arm" (i.e., a simulation of an arm rather than an actual physical implementation).

Nengo is not adequately described in their publications for us to make an evaluation of it. As a software system being run on what was reported to be an "eight core Xeon conventional computer,"[3] it is certain that the execution of Eliasmith's program did not run in real time. Because their goal was primarily conceptual and cognitive neuroscientific, this is hardly an objection to their work. However, the Eliasmith project can be criticized in the context that they set for themselves—linking the "incredible" complexities of brain and behavior. It was not made clear in their publication how this work actually accomplishes that task.

In sum, Spaun is another programmed (as opposed to hardware) instantiation of a system of 2.5 million simulated neurons that its authors claim has the ability to carry out what are analogs of cognitive activity. It is of special interest in the present context because its motivation arises from the cognitive neuroscience task rather than from engineering or neurophysiological considerations.

The European Human Brain Project

Perhaps the most ambitious project currently aimed at developing a supercomputer model of the brain is the European Human Brain Project (HBP). This project has, as its ultimate goal, a computer simulation of the entire human brain (Markram, 2006), ranging from the biochemical level of ionic distributions to the whole brain with all of its neurons and intricate synaptic connections. A complete simulation of this magnitude would require the parallel interaction of almost 100 billion neurons and would probably require at least an ExaFLOPS-capable computer—a device that does not yet exist. Current work is being carried out on an IBM Blue Gene computer capable of simulating about 200 million neurons.

The strategy used by the HBP to achieve this goal is to develop a succession of ever more complex substructures of the brain, including a standard model of a single neuron that closely approximates neurophysiological reality. Some may argue that the details of the molecular properties of neurons are not particularly relevant to explication of the information-processing network properties germane to the process of cognitive neuroscience. However, in principle, this is a brain project, not one aimed at the mind–brain question. Currently, it has been reported that the HBP has been successful in developing a model for a cortical column in the cerebral cortex of a rat—a structure that contains about 10,000 neurons—and an aggregation of 100 of these simulated columns into a single program for a total of about a million simulated neurons. However, it has proven difficult to find fully vetted publications describing the accomplishments of the HBP in detail; the most recent serious formal publication seems to have been that one written in 2006. Other reports are essentially ambitious projections, proposals, or incomplete statements of long-term goals.

The HBP is not intended at this stage of the game to be a model of cognitive processes; it is an effort whose goal is to simulate a complex system—the brain—at all levels of neural activity starting from the basic biochemistry of neurons, moving on to interacting neurons and synapses, and, in the long-term future, arriving at some connections to cognition. Therefore, it does not address the mind-body problem directly; nevertheless, if it should succeed, ambitious simulation of this level would massively contribute to our understanding of how cognitive processes might be encoded at all neural levels.

Unfortunately, despite this promise, large-scale simulations such as the HBP have come under severe criticisms in recent years by neuroscientists for several reasons. Among the most important is the problem of controlling and displaying the input stimuli and the output response of such a system. Determining the initial conditions (the starting states of each and every one of the multitude of neurons simulated in the supercomputer) would be a formidable job unto itself. Obviously, some novel display methodologies similar to those developed by Izhikevich and Edelman would be necessary but those plots of neuronal activity versus time are very close to the raw data and more imaginative display techniques will be needed. Determining the output (i.e., the final states) of millions, not to say the billions of simulated neurons becoming possible with today's very large-scale computers would also strain current display technology unless in some unexpected way the simulation communicated the musings of an independent sentience.[4]

In 2014, an open letter signed by 776 mainly European neuroscientists was sent to the European Commission, the organization that funded the HBP (to the prodigious and unprecedented promised amount of a billion euros) in which it was stated that "HBP is not a well-conceived or implemented project and that it is ill suited to be the center piece of European neuroscience" (from Summary of Open Message to the European Commission concerning the Human Brain Project, 18 July 2014). The project has been characterized as being "opaque," "poorly managed," and "premature" and, although intended to be a collaboration of more than a hundred universities, has suffered from a loss of many of the institutions that were originally involved in planning the project. Others have challenged its goals, arguing that it is actually a computer development project hiding within the "sheep's clothing" of brain research.[5]

The HBP is an extremely controversial approach whose goals are, to a significant degree, more hyperbole than scientific accomplishment as evidenced in the absence of any refereed research publications. The latest development in this ongoing tale is that the billion-euro funding from the European Union has been redirected and "a re-evaluation of the scientific goals that relegate modeling of the whole brain to the back burner" (Enserink, 2015).

In sum, the HBP is aimed at a complete simulation of neurophysiological networks from the level of membrane dynamics to cognitive processes. It is currently being run on an IBM Blue Gene computer. The project is reported to have programmed a neuronal network that is the equivalent of a cortical column consisting of 10,000 simulated neurons and 10^8 synapses.

This is an ambitious, perhaps overly ambitious, project that it is hoped will offer insights into network organization. Currently the organizers of this project seem to have little interest in bridging the gap to cognition.

The United States BRAIN Project

A comparably ambitious, but hopefully more modest in its goals, research program is now being financed by the United States government along with a number of private contributors. The goal of this effort (Brain Research through Advancing Innovative Neurotechnologies [BRAIN]) is to develop methodologies for simultaneously recording and stimulating activity in multiple neurons. Although not as extravagantly funded and much more diffuse in its goals than the HBP, BRAIN has also been criticized as putting the "cart before the horse," that is, developing special equipment before the need for it has been demonstrated.

The DARPA SyNAPSE Project

In 2008, the Defense Advanced Research Projects Agency (DARPA) initiated and funded a program to develop a new computer architecture based on the parallelicity idea. The motivating assumption was that the goals of this project could be best achieved by imitating some of the features of the organic brain. This project, which continues to this day, is referred to as the Systems of Neuromorphic Adaptive Plastic Scalable Electronics (SyNAPSE) program. The goals of this project were to develop new "neuromorphic" logical circuits based on the properties of biological neurons as well as to develop parallel processing computer systems consisting of novel hardware, programming, and system architecture. Secondary goals included engineering goals such as power minimization and redundant architecture that were resistant to local faults. The original participants in the SyNAPSE program included a large number of universities divided into two teams headed by IBM and HRL (an off-shoot of Hughes Research Laboratory), respectively. Funding now approaches $100 million.

The two laboratories have taken different directions in the latter years of the SyNAPSE funding cycle. HRL's goal has been to concentrate on system architectures that behave very much as do neurons; specifically self-programming, learning, and what we would call today habituation. IBM has concentrated on building very powerful circuit boards that can be grouped together to build ever larger parallel processing computers. HRL's part of the project has emphasized biologically realistic properties of synapses. Thus, their circuits are smaller and designed to simulate processes that affect the strength of synaptic conductivity such as those suspected to be involved in learning and habituation. HRL concentrated on developing practical applications of devices that can learn to improve their performance. An autonomous drone helicopter is one of their current successes—the most recent version is controlled by a single chip consisting of 576 simulated neurons.

Over the past decade, the IBM part of the project has gone in a different direction—they have aimed at developing novel computer architectures with ever higher numbers of cores with faster and faster processing capability. Using the Sequoia supercomputer, one of the most powerful parallel processing supercomputers available in 2012, the IBM group was able to execute a program that they claim simulated a system of 500 billion neurons and 100 trillion synapses (as described in the article by Service, 2014). This had to be done with an iterative process since the number of actual cores was only 1.5 million and the logic was built on relatively slow conventional integrated circuits. As a result, the program ran much more slowly than did estimates of the parallel processing speed of the human brain. The Sequoia project was not designed to simulate the human brain; it was mainly intended to develop an efficient computer hardware system that was "inspired" by the brain.

The next step in the development of such a system was to engineer a unitized "neuromorphic" large-scale integrated circuit chip that incorporated the processing features of many neurons by means of specialized processing cores. The IBM TrueNorth device that emerged from these primarily engineering goals was a single chip that contains 5.4 billion transistors and 256 million simulated synapses simulating the behavior of a million simulated neurons. The logic design was organized in a manner that was assumed to be comparable to the complexities of the human brain. Current research reports suggest that it has been possible to link "more than a dozen" of these integrated circuit chips together, but the developers feel that this is only the beginning and that the system is capable of unlimited upwards scalability by simply attaching more and more of the TrueNorth chips to one another.

One of the most interesting aspects of the TrueNorth project is that the cores on these chips are not based on the serial von Neumann prototype. That classical design placed the core and the memory in different locations, resulting in high-energy demand whenever information was transmitted between the memory and the cores. This limitation of the von Neumann–type core could be abetted by placing memory and cores on the same chip, thus reducing transmission times and resulting in a remarkably low power demand. The IBM group (Merolla et al., 2014) reports that the TrueNorth chip drew only 63 milliwatts.

The development of the TrueNorth chip was both an important milestone in packing circuits onto a single chip and novel logic design. However, it consisted of only a relatively small number of simulated cores and neurons in comparison to the 2012 Sequoia-based project. What it accomplished should not be confused with the kind of progress made in the 2012 project. The 2012 project carried out programs on a supercomputer that simulated numbers that actually exceeded the biological brain's neuron counts (530 billion) but had to do so by repetitive use of the circuitry in the Sequoia computer they utilized and thus not in real time. The ultimate goal of the TrueNorth project, on the other hand, is to design circuitry that mimics these large numbers of neurons in the brain and pack it onto a single chip—a chip that could be scaled upwards by interconnection with other TrueNorth chips to represent billions of neurons. The TrueNorth chip has properties that may ultimately achieve this goal, but it has not yet been expanded to approximate the number of neurons in the brain—86 billion.

The IBM TrueNorth project, although motivated mainly by engineering goals, also has been used to simulate some relatively simple cognitive processes. Inputs were exemplars of a set of moving objects. The behavioral goal of the system was correct detection and recognition by the TrueNorth chip of these different objects; however, no performance data were provided in the brief articles describing the TrueNorth-based simulations nor were the methods used to input the initial conditions or to develop program modification (i.e., learning) as a result of the activity in the chip.

In summary, the HBP and the TrueNorth projects had many similarities in their goals; however, they also were quite different in some ways. Specifically, the main goal of the HBP is to program whatever computer system is available to simulate billions of neurons in the human brain and to do this at multiple levels of analysis. As each level is simulated, the intent is to concatenate these various simulations until a neural "model" of the brain emerges. The target in this case is the neuroanatomy and neurophysiology of the brain itself.

The IBM work is, on the other hand and quite to the contrary, aimed at the development of hardware, inspired to a degree by the biology of the brain, but more concerned with such functional properties as power efficiency, fault tolerance, packing density, and scalability, as well as programmability. The brain, to the IBM group, is only a distant benchmark to be implemented someday by complex circuitry—the simple behavioral functions are merely a means of demonstrating the capabilities of the system. Should there be any insights about the brain and specifically about what artificial cognitive functions are obtained during this engineering-motivated development project, that would be "icing on the cake." In the meantime, allusions to the brain-like features of all of these supercomputers might well be taken with extreme caution.

5.4 Interim Summary

There is little question that the large-scale supercomputer represents the most promising potential breakthrough tool in the search for a solution to the mind–brain problem. The foundation premises of the two systems—the human brain and the modern computer—are the same. Each is based on the general principle of a network of nodes, cores, or neurons that is able to carry out many computations simultaneously by virtue of its ability to parallel process. Furthermore, the supercomputers that are available today (or will be soon) are approaching the number of neurons in the organic brain. Although the promise is great, none of these large-scale projects, regardless of their biological or engineering goals, have yet achieved anything close to functional equality with the brain.

Although liberally sown with the vocabulary of psychology and neurophysiology, there is nothing that appears to be even the harbinger of a theory of mind–brain relations in the published literature. Indeed, we have not yet learned how to carry out a relevant experiment on these supercomputers. A major problem inhibiting such accomplishments is the difficulty of programming these computational behemoths. In actuality, the task is comparable to the analogous development process that occurs at least over the lifetime of an individual human and possibly over eons of organic evolution. Indeed, it is not even sure what the word "programming" would mean in the context of a computer that operates in parallel on many components of the input and responds in the form of a distributed network.

In large part, whatever the actual motivations and goals, many supercomputer engineering projects are being promoted on the basis of promises and hopes in the fields of neurophysiology and psychology. Although the power of the brain is clearly a function of the myriad synaptic connections, we still have no idea what are the specific patterns of these connections that correspond to behavior selection, much less consciousness. Without question, newer and faster computer systems are going to emerge from this activity, but we are still in need of a conceptual revolution to determine how the parallel processing supercomputer will contribute to cognitive neuroscience.

Surveying this chapter, we may draw a number of tentative and preliminary conclusions about the promising role of these new computers as theory engines:

1. The development in recent years of the computer from a simple single-core system to one with billions of functional components has been amazing.
2. The simple concept of being able to store both data and instructions in the computer memory was a key factor in the development of the modern stored program computer.
3. Another key factor was the ability to modify the program during its execution.
4. Speeding up a computer simply by adding processing nodes is not always possible. Some programs are not inherently decomposable; in this case Amdhal's law holds wherein the speedup in processing is equal to 1/1–P where P is the proportion of the program that can be parallelized.
5. Conceptually, the parallel processing supercomputer promises to come closer than any other available tool or methodology to understanding the information-processing aspects of the brain.
6. A challenging conceptual and technological barrier to theory development is the fact that we do not yet know how to interpret the information processes that may be carried out by these systems. Specifically it is not yet clear how we would program them to set up initial states, nor do we know how to interpret the patterns of activity that might correspond to organic mental activity.
7. A related issue is that we do not have a satisfactory way of defining and measuring cognitive processes given their inaccessibility. How would we know that a powerful computer is conscious? The well-known Turing test, despite popular opinion, does not distinguish the conscious from the automatic.

8. We do not yet have a solid idea of how many neurons in the brain are required to carry out cognitive processes or how many cores are necessary to solve a computer problem.
9. Large-scale parallel processing computer research is being carried out with multiple motivations and goals. Most are aimed at engineering development or simulation of particular neuronal networks. Often, research in this field is embellished with words such as "brain" or "neuron," but the cognitive and neurophysiological meanings of these words is long lost. Many such "brain" projects have little to do with the goals of cognitive neuroscience.
10. Looming on the horizon is what is probably the most significant practical obstacle of all—the combinatorial explosion. The enormity of combinations and possibilities is not yet a part of the thinking of most psychologists and physiologists. There may well be things we cannot do because it would require computers as large as the knowable universe.
11. Similarly, the criteria for demonstrating equivalence of brain function and cognitive processes are far more stringent than generally understood. How will we know when a simulation provides a necessary explanation and not just a sufficient one?
12. A sampling of the large-scale parallel processing simulations being carried out in recent years displays the diversity of their goals. Some are strictly engineering, some are neurophysiological, and a few aspire to being models of cognition. Most end up being examples of cumulative neuronal activity that are not linked to cognition. Some end up being restricted to sensory processes rather than anything that might be called cognitive. Some, after years of research and huge amounts of money expended, end up with nothing of relevance to cognitive neuroscience. The words used to describe one's favorite project often do not discriminate between these disparate goals and findings.

In conclusion, the advent of large-scale parallel processing supercomputers has opened up a whole new approach to the problem of how we deal with the huge number of neurons and synapses populating the organic brain. Many imponderables and challenges still lie ahead before we can assert that we have determined the codes for cognitive process; however, this approach is in its infancy and only the future can tell if the Hebbian concept of interacting neurons represents a tractable or intractable path to solving the most important problem of human history.

Notes

1. It is important to remember that not all "neural" computers engage ideas from neurophysiology. The mathematics specific to neural-like networks is characterized by a goal of dealing with networks of linear threshold units. Engineers are concerned with increasing the speed and power of computers using strategies that may range from simple metaphors of "brain-like" networks without any specific interest in building a

theory of cognition. In this book, I have striven, not always successfully, to find examples of projects that have specific cognitive neuroscience implications.

2. About this time, I (Uttal, 1975) proposed an autocorrelation theory of form detection that was also a parallel processing system. Calculations of the model were successful in predicting the detectability of dotted forms hidden in dotted noise.
3. The Xeon computer CPU is a modestly powered core used in a system that is just barely above a simple desk computer. It is not possible to estimate the speed at which such a computer might run a 2.5 million-neuron simulation. However, it is very clear that, even with eight cores, it would not be comparable in execution time to any of the high-level parallel processors discussed elsewhere.
4. Some glimmering of hope for the development of super displays is on the horizon. Ahrens et al. (2013), for example, have developed a technique combining optogenetic and optical techniques to image the entire body of a larval zebrafish. This permitted them to simultaneously show 80% of neurons in this model preparation at a resolution that allowed them to scrutinize individual neurons. Although promising a considerable technical accomplishment, this technique does not solve the problem of numerousness (the data explosion) nor of complexity (the combinatorial explosion).
5. Although the same criticism may be made of the United States–sponsored BRAIN project, that very large project has much more modest goals—namely to provide the instruments necessary to "reverse engineer" the brain at some future time.

6

EMERGING CONCLUSIONS

This book has been aimed at partially answering one practical question and one theoretical question. The practical question is—how does the available technology affect microneuronal theory development in cognitive neuroscience? To answer the first question, it has been deemed to be appropriate to survey the nature of the various kinds of instruments that are available, the kinds of empirical results that have been forthcoming from their application, and the relevant theories. It is clear that remarkable progress has been made in both instrumentation and empirical neurophysiological research and that the impact on contemporary theory has been substantial.

The main answer to the second question—the theoretical one—that we must draw from the review carried out in this book is that despite substantial progress in understanding the brain, we still have no idea of how the mental emerges from the physiological. What we have, instead, is an unsubstantiated but increasing appreciation that the key to solving the theoretical question is going to be found at the microneuronal level, that is, at the level at which the properties of the individual neuron or of networks of neurons are preserved. It is not to be found, as I argued in my companion book (Uttal, 2016), at the macroneural level that characterizes so much of current brain-imaging research. It is not, I argued there, that the cumulative, statistically summarized, pooled macro-responses that purportedly represent the codes for cognition, but, instead, the nature of individual neurons and their combinations and interactions.

This book has surveyed the technology of microneuronal research and the kinds of theories that are driven by observations that are obtained at this microneuronal level. It is a truism that none of these methods is perfect and that each is flawed in some fundamental manner or is still in an early stage of development. Nevertheless, much of what we obtained at the level of single neurons is based on some outstanding accomplishments in measurement of the activity of these microscopic entities. The fortuitous availability of model preparations has in some cases permitted us to study some relatively simple neuronal networks. Unfortunately, baldly extrapolating these essentially single-neuron neurophysiological accomplishments to theory leads one to one of the great dead ends of modern mind–brain theory—the idea that single neurons can encode or represent complex cognitive concepts. Equally unfortunately, our ability to examine even the simplest of networks is limited to structures involving only a few neurons.

The most promising new technology capable of influencing theory now appearing on the scene is the large-scale parallel processing "supercomputer." This kind of information-processing giant has now been engineered to a point that, in principle at least, it should be capable of enlisting a vast number of computing elements, a number that is comparable to the number of neurons in the human brain. As powerful as it is, however, we still have only the most primitive ideas concerning how to run a computer experiment involving so many components—how, for example, do we initialize such a behemoth and how would we know what it is doing.

The overarching theoretical question, therefore, remains unanswered. Worst of all, we do not yet even know if the mind–brain problem is solvable. Over the years, I have struggled with this general question without resolution. The differences between the properties of the tangible brain and the intangible mind are so great that the whole approach to seeking a neuroreductionist answer to the question of the nature of the mind may be an impossible question or a bad question.

Regardless whether we will ultimately be able to answer the grand question, there are many less earth-shaking issues that can be studied and from which general understanding and formal, if specialized, theories will eventually emerge. Whether it be potential progress in understanding networks or better descriptions of cognitive processes, there is much to be learned about the nature of mind–brain relations before we abandon the chase. The techniques and theories described here go hand-in-hand and the mutual impact that each has on the other should not be underestimated. No one who is interested in the field of cognitive neuroscience can survive professionally in the future if unacquainted with these techniques and their implications. If this book introduces both practice and theory to those who have aspirations in this field, then it will have served its purpose.

In this last chapter, I present a list of emerging observations and principles that summarize the extended discussions of the preceding chapters. Many of these statements, admittedly, are subject to controversy and varied interpretation. However, I believe collectively that they represent a comprehensive statement about the current state of the microneural approach to cognitive neuroscience.

1. The most fundamental postulate in cognitive neuroscience is what we may call the ontological assumption. This postulate asserts that all cognitive processes in some unknown way are accounted for in terms of the neurophysical processes and organization of the brain. Whatever it is that will ultimately explain the mind–brain relation, it cannot in any way countervail the laws of physics. Without this postulated bridge between mind and matter the enterprise of cognitive neuroscience could not exist.
2. The ultimate theoretical goal of this science is to determine the "psychoneural equivalent" of cognition, that is, to determine which neural mechanisms embody which cognitive processes.
3. The mind–brain problem is enormously complex, and it is not yet certain whether it is scientifically tractable.
4. Despite this difficulty, theories abound in cognitive neuroscience in which mental activity is associated with brain components and functions. These theories exist at many different levels of analysis.
5. This book has concentrated on the microlevel—the level of neurons and synapses—and of networks of them.
6. A basic assumption in writing this book was that the methodology and equipment used at any time in history strongly determined the type of theory that will emerge despite what may be ample countervailing evidence.
7. Theories can be categorized as reductive and nonreductive. Reductive theories may be neural or cognitive depending on the elements from which they are supposed to be constructed. Nonreductive or descriptive theories track the trajectory of a process but do not attempt to analyze the underlying mechanisms.
8. Theories can also be categorized as top-down or bottom-up. The former is the strategy used by psychologists and other behavioral scientists. The latter is the approach taken by neurophysiologists.
9. Currently, the most promising path to an overarching theory of mind–brain relationships lies at the microscopic network level in which neuronal organization is the key. Top-down theories can never be definitive because of the underdetermined nature of critical evidence.
10. Single cell theories are deeply flawed and implausible at their conceptual roots. Although there is no question that some neurons in the brain respond highly selectively to specific stimuli, the converse is questionable. That is, activation of a single neuron to determine what its cognitive effects are is an undoable experiment. Single neuron theories will always be based on Gedanken experiments and analogies rather than actual ones.
11. Even if we could carry out an experiment, we are confronted with the "exclusivity" problem. That is, it is impossible to know what effect an individual neuron is having on mental processes in the cacophony of all of the other responding neurons. In other words, how does one of many neurons produce or dominate the cognitive outcome to the exclusion of all others?

12. The idea of a network of many interacting neurons has begun to dominate and replace theoretical thinking in cognitive neuroscience. This approach was stimulated by the pioneering work of Donald Hebb, who postulated networks of neurons organized into functional groups whose configurations varied with the state of the involved synaptic junctions.
13. Unfortunately, the Hebb conjecture is probably untestable since it requires enormous amounts of information about the detailed structure of the network. No conceivable neurophysiological methodology permits experiments of this kind in which either adequate data are accumulated or stimulus configurations are properly controlled.
14. Some (how many can only be guessed at) cognitive neuroscientists seem not to appreciate the basic principles of combinatorial arithmetic. A very small sample of nodes (neurons) can produce astronomical numbers of combinations—numbers that often turn out to exceed the space available in the known universe or the durations that exceed the known age of the universe. The possibility of an exhaustive analysis of a system like the brain is remote, perhaps even demonstrably impossible. Even such superficially simple tasks such as spike action potential clustering or the determination of functional connectivity maps may be impossible because of combinatoric constraints.
15. The direct neurophysiological approach in which the arrangement and states of the neurons involved in a real neuronal network is also unlikely to be achieved. It is not possible in even the most reduced model preparation to track out the functions of a network consisting of more than a few neurons.
16. Furthermore, it is never certain whether a mechanism observed in a model nervous system is implementing the same function in the same way as the ones underlying human thought. Indeed, it is more likely in a simplified model preparation that we are looking at analogs rather than homologs of the processes of interest.
17. Redundancy is a basic fact of life in the nervous system. An observed neurophysiological mechanism may be only one of many alternative (and constantly shifting) structures.
18. The parallel processing supercomputer is achieving numbers of synthetic or programmed neurons that are so great that it is feasible to begin to think of the possibility of simulating human brains. However, loading a supercomputer with initial conditions and recovering the myriad details of the resulting response remain major difficulties. However, this supercomputer approach is only a few years old and much is to be learned about what they can and cannot offer to the problems of cognitive neuroscience theory development.
19. Even a successful simulation of a cognitive process on a supercomputer would not solve the problem of distinguishing between a sufficient and clever analog, on the one hand, and a correct and necessary explanatory explanation

of how the magical transformation of tangible brain into intangible mind occurs, on the other. A functional simulation is not an explanation.

20. Despite the many obstacles and uncertainties confronting research in cognitive neuroscience, there is no more exciting and significant scientific problem than the mind–brain one. I have no regrets about spending my career on it, no matter how impenetrable the task seems at the moment.

BIBLIOGRAPHY

Adamos, D. A., Kosmidis, E. K., & Neophilidis, G. (2008). Performance evaluation of PCA-based spike sorting algorithms. *Computer Methods and Programs in Biomedicine, 91*(3), 232–244.

Ahrens, M. B., Orger, M. B., Robson, D. N., Li, J. M., & Keller, P. J. (2013). Whole-brain functional imaging at cellular resolution using light-sheet microscopy. *Nature Methods, 10*, 413–420.

Alivisatos, A. P., Chun, M. Y., Church, G. M., Deisseroth, K., Donoghue, J. P., Greenspan, R. J., . . . Yuste, R. (2013). The brain activity map. *Science, 339*(6125), 1284–1285.

Amdahl, G. M. (1967). *Validity of the single processor approach to achieving large-scale computing capabilities.* Paper presented at the AFIPS Spring Joint Computer Conference, April 1967, Atlantic City, NJ.

Ananthanarayanan, R., Esser, S. K., Simon, H. D., & Modha, D. S. (2009). *The cat is out of the bag: Cortical simulations with 10^9 neurons, 10^13 synapses.* Paper presented at the Proceedings of the Conference on High Performance Computing Networks, Storage and Analysis, SC '09, New York, NY.

Anderson, J. A., Pellionisz, A., & Rosenfeld, E. (1993). *Neurocomputing II.* Cambridge, MA: MIT Press.

Anderson, M. L. (2010). Neural reuse: A fundamental organizational principle of the brain. *Behavioral and brain sciences, 33*(04), 245–266.

Argonov, V. Y. (2012). Neural correlate of consciousness in a single electron: Radical answer to "quantum theories of consciousness." *Neuroquantology, 10*(2), 276–285.

Aue, T., Lavelle, L. A., Cacioppo J. T. (2009). Great expectations: What can fMRI research tell us about psychological phenomena? *International Journal of Psychophysiology, 73*, 10–16.

Azevedo, F. A. C., Carvalho, L. R. B., Grinberg, L. T., Farfel, J. M., Ferretti, R. E. L., Leite, R. E. P., . . . Herculano-Houzel, S. (2009). Equal numbers of neuronal and nonneuronal cells make the human brain an isometrically scaled-up primate brain. *Journal of Comparative Neurology, 513*(5), 532–541.

Bakker, M., & Wicherts, J. M. (2011). The (mis)reporting of statistical results in psychology journals. *Behavior Research Methods, 43*, 666–676.

Bandettini, P. (2007). Functional MRI today. *International Journal of Psychophysiology, 63*(2), 138–145.

Bargmann, C. I., Hartwieg, E., & Horvitz, H. R. (1993). Odorant-selective genes and neurons mediate olfaction in C. elegans. *Cell, 74*(3), 515–527.

Barlow, H. B. (1953). Summation and inhibition in the frog's retina. *Journal of Physiology-London, 119*(1), 69–88.

Barlow, H. B. (1961). Possible principles underlying the transformations of sensory messages. In W. A. Rosenblith (Ed.), *Sensory Communication* (pp. 217–234). Cambridge, MA: MIT Press and New York: John Wiley.

Barlow, H. B. (1972). Single units and sensation: A neuron doctrine for perceptual psychology. *Perception, 1*, 371–394.

Barrett, L. F. (2011). Bridging token identity theory and supervenience theory through psychological construction. *Psychological Inquiry, 22*, 115–127.

Barrett, L. F. "What Emotions Are (and Aren't)." *The New York Times*, August 2, 2015.

Barrett, L. F., & Satpute, A. (2013). Large-scale brain networks in affective and social neuroscience: Towards an integrative architecture of the human brain. *Current Opinion in Neurobiology, 23*, 361–372.

Bashir, S., Perez, J. M., Horvath, J. C., Pena-Gomez, C., Vernet, M., Capia, A., . . . Pascual-Leone, A. (2014). Differential effects of motor cortical excitability and plasticity in young and old individuals: A Transcranial Magnetic Stimulation (TMS) study. *Frontiers in Aging Neuroscience, 6*(111), 1–13.

Becker, D. J. (1995). *Beowulf: A parallel workstation for scientific computation.* Paper presented at the International Conference for Parallel Processing.

Belliveau, J. W., Kennedy, D. N., Mckinstry, R. C., Buchbinder, B. R., Weisskoff, R. M., Cohen, M. S., . . . Rosen, B. R. (1991). Functional mapping of the human visual-cortex by magnetic-resonance-imaging. *Science, 254*(5032), 716–719.

Belykh, I., de Lange, E., & Hasler, M. (2005). Synchronization of bursting neurons: What matters in the network topology. *Physical Review Letters, 94*(18).

Benjamin, B., Gao, P., McQuinn, E., Choudhary, S., Chandrasekaran, A. R., Bussat, J. M., . . . Boahen, K. (2014). Neurogrid: A mixed-analog-digital multichip system for large-scale neural simulations. *Proceedings of the IEEE, 102*(5), 699–716.

Bennett, M. R. (2001). *History of the Synapse.* Boca Raton, FL: CRC Press.

Berger, H. (1929). Uber das electroenkephalogramm de menschem. *Archiv fur psychiatric und Nervenkrankheiten, 97*, 527–570.

Binet, A. (1907). *The Mind and the Brain.* London: K. Paul, Trench, Trübner & Co. Ltd.

Blake, A. J., Rodgers, F. C., Bassuener, A., Hippensteel, J. A., Pearce, T. M., Pearce, T. R., . . . Williams, J. C. (2010). A microfluidic brain slice perfusion chamber for multisite recording using penetrating electrodes. *Journal of Neuroscience Methods, 189*(1), 5–13.

Boahen, K. (2006). Neurogrid: Emulating a million neurons in the cortex. *Conference Proceeding of the IEEE Engineering in Medicine and Biology Society, Supplement*, 6702.

Bowers, J. S. (2009). On the biological plausibility of grandmother cells: Implications for neural network theories in psychology and neuroscience. *Psychological Review, 116*(1), 220–251.

Breasted, J. H. (1991). *The Edwin Smith Surgical Papyrus* (Vol. 3–4). Chicago: University of Chicago Oriental Institute Publications.

Brindley, G. S. & Merton, P. A. (1960). The absence of position sense in the human eye. *The Journal of Physiology, 153*, 127–130.

Brown, E. N., Kass, R. E., & Mitra, P. P. (2004). Multiple neural spike train data analysis: State-of-the-art and future challenges. *Nature Neuroscience*, 7(5), 456–461.

Burks, A. R., & Burks, A. W. (1989). *The First Electronic Computer: The Atanasoff Story*. Ann Arbor: University of Michigan Press.

Button, K. S., Ioannidis, J. P. A., Mokrysz, C., Nosek, B. A., Flint, J., Robinson, E. S. J. & Munafò, M. (2013). Power failure: Why small sample size undermines the reliability of neuroscience. *Nature Reviews Neuroscience, 14*, 365–376.

Cajal, S. R. Y. (1906, December 12). *The structure and functions of neurons.* Paper presented at the Nobel Foundation Award Ceremony. Stockholm, Sweden.

Carew, T. J., Hawkins, R. D., & Kandel, E. R. (1983). Differential classical-conditioning of a defensive withdrawal reflex in Aplysia-Californica. *Science, 219*(4583), 397–400.

Carew, T. J., Pinsker, H. M., & Kandel, E. R. (1972). Long-term habituation of a defensive withdrawal reflex in Aplysia. *Science, 175*(4020), 451–454.

Cariani, P. A. (2004). Temporal codes and computations for sensory representation and scene analysis. *IEEE Transactions on Neural Networks, 15*(5), 1100–1111.

Carp, J. (2012a). The secret lives of experiments: Methods reporting in the fMRI literature. *NeuroImage, 63*(1), 289–300.

Carp, J. (2012b). On the plurality of (methodological) worlds: Estimating the analytic flexibility of fMRI experiments. *Frontiers in Neuroscience, 6*, 149.

Carpenter, G. A., Grossberg, S., & Reynolds, J. H. (1991). Artmap—Supervised real-time learning and classification of nonstationary data by a self-organizing neural network. *Neural Networks, 4*(5), 565–588.

Castellucci, V., & Kandel, E. R. (1976). Presynaptic facilitation as a mechanism for behavioral sensitization in Aplysia. *Science, 194*(4270), 1176–1178.

Castellucci, V., Pinsker, H., Kupferma, I., & Kandel, E. R. (1970). Neuronal mechanisms of habituation and dishabituation of gill-withdrawal reflex in Aplysia. *Science, 167*(3926), 1745–1748.

Chalmers, D. J. (1996). *The Conscious Mind: In Search of a Fundamental Theory*. New York: Oxford University Press.

Charkhkar, H., Knaack, G. L., Gnade, B. E., Keefer, E. W., & Pancrazio, J. J. (2012). Development and demonstration of a disposable low-cost microelectrode array for cultured neuronal network recording. *Sensors and Actuators B-Chemical, 161*(1), 655–660.

Chiang, A. S., Lin, C. Y., Chuang, C. C., Chang, H. M., Hsieh, C. H., Yeh, C. W., . . . Hwang, J. K. (2011). Three-dimensional reconstruction of brain-wide wiring networks in Drosophila at single-cell resolution. *Current Biology, 21*(1), 1–11.

Christofides, N. (1975). *Graph Theory: An Algorithmic Approach*. New York: Academic Press.

Cogan, S. F. (2008). Neural stimulation and recording electrodes. *Annual Review of Biomedical Engineering, 10*, 275–309.

Cohen, T. E., Kaplan, S. W., Kandel, E. R., & Hawkins, R. D. (1997). A simplified preparation for relating cellular events to behavior: Mechanisms contributing to habituation, dishabituation, and sensitization of the Aplysia gill-withdrawal reflex. *Journal of Neuroscience, 17*(8), 2886–2899.

Coltheart, M. (2006). What has functional neuroimaging told us about the mind (so far)? *Cortex, 42*(3), 323–331.

Cox, D. R., & Smith, W. L. (1953). The superposition of several strictly periodic sequences of events. *Biometrika, 40,* 1–11.

Cox, D. R., & Smith, W. L. (1954). On the superposition of renewal processes. *Biometrika, 41*(1–2), 91–99.

Crick, F., & Koch, C. (1995). Are we aware of neural activity in primary visual-cortex? *Nature, 375*(6527), 121–123.

Crick, F., & Koch, C. (1998). Consciousness and neuroscience. *Cerebral Cortex, 8*(2), 97–107.

Cullington, H. E., & Zeng, F. G. (2008). Speech recognition with varying numbers and types of competing talkers by normal-hearing, cochlear-implant, and implant simulation subjects. *Journal of the Acoustical Society of America, 123*(1), 450–461.

Damadian, R. (1971). Tumor detection by nuclear magnetic resonance. *Science, 171*(3976), 1151–1153.

Dawson, G. D., & Scott, J. W. (1949). The recording of nerve action potentials through skin in man. *Journal of Neurology, Neurosurgery, and Psychiatry, 12,* 259–267.

de Garis, H., Chen, S., Goertzel, B., & Lian, R. T. (2010). A world survey of artificial brain projects, Part I: Large-scale brain simulations. *Neurocomputing, 74*(1–3), 3–29.

DeMonasterio, F. M., & Gouras, P. (1975). Functional properties of ganglion-cells of rhesus-monkey retina. *Journal of Physiology-London, 251*(1), 167–195.

Desai, S., Stickney, G., & Zeng, F. G. (2008). Auditory-visual speech perception in normal-hearing and cochlear-implant listeners. *Journal of the Acoustical Society of America, 123*(1), 428–440.

DeWeerth, S. P., Meacham, K. W., Giuly, R. J., Guo, L., & Hochman, S. (2008). A lithographically-patterned, elastic multi-electrode array for surface stimulation of the spinal cord. *Biomedical Microdevices, 10*(2), 259–269.

di Pellegrino, G., Fadiga, L., Fogassi, L., Gallese, V., & Rizzolatti, G. (1992). Understanding motor events—A neurophysiological study. *Experimental Brain Research, 91*(1), 176–180.

Djourno, A., & Eyries, C. (1957). Prothese auditive par excitation electrique a distance du nerf sensoriel a laide dun bobinage inclus a demeure. *Presse Medicale, 65*(63), 1417–1417.

Duhem, P. (1914). *La Theorie Physique: Son Objet et Sa Structure.* Paris: Marcel Riviera & Cie.

Ekanadham, C., Tranchina, D., & Simoncelli, E. P. (2014). A unified framework and method for automatic neural spike identification. *Journal of Neuroscience Methods, 222,* 47–55.

Eliasmith, C., Stewart, T. C., Choo, X., Bekolay, T., DeWolf, T., Tang, Y., Rasmussen, D. (2012). A large-scale model of the functioning brain. *Science, 338*(6111):1202–5. doi:10.1126/science.1225266

Enroth-Cugell, C., & Robson, J. G. (1966). The contrast sensitivity of retinal ganglion cells of the cat. *Journal of Physiology-London, 187*(3), 517–552.

Enserink, M. (2015). A [Euro] billion brain reboot. *Science, 347*(6229), 1406–1407.

Erickson, J., Tooker, A., Tai, Y. C., & Pine, J. (2008). Caged neuron MEA: A system for long-term investigation of cultured neural network connectivity. *Journal of Neuroscience Methods, 175*(1), 1–16.

Eversmann, B., Jenkner, M., Hofmann, F., Paulus, C., Brederlow, R., Holzapfl, B., . . . Thewes, R. (2003). A 128x128 CMOS biosensor array for extracellular recording of neural activity. *IEEE Journal of Solid-State Circuits, 38*(12), 2306–2317.

Farah, M. J. (2014). Brain images, babies, and bathwater: Critiquing critiques of functional neuroimaging. *Hastings Center Report, 44*, S19–S30.

Fatt, P., & Katz, B. (1952). Spontaneous subthreshold activity at motor nerve endings. *Journal of Physiology-London, 117*(1), 109–128.

Feigl, H. (1958). The mental and the physical. *Minnesota Studies in the Philosophy of Science, 2*, 370–497.

Field, D. J. (1994). What is the goal of sensory coding? *Neural Computation, 6*(4), 559–601.

Flood, T. F., Gorczyca, M., White, B. H., Ito, K., & Yoshihara, M. (2013). A large-scale behavioral screen to identify neurons controlling motor programs in the Drosophila brain. *G3-Genes Genomes Genetics, 3*(10), 1629–1637.

Flynn, M. J. (1972). Some computer organizations and their effectiveness. *IEEE Transactions on Computers, C-21*(9), 948–960.

Fodor, J. A. & Pylyshyn, Z. W. (1988). Connectionism and cognitive architecture: A critical analysis. *Cognition, 28*, 3–71.

Fox, M. D., Snyder, A. Z., Vincent, J. L., Corbetta, M., Van Essen, D. C., & Raichle, M. E. (2005). The human brain is intrinsically organized into dynamic, anticorrelated functional networks. *Proceedings of the National Academy of Sciences of the United States of America, 102*(27), 9673–9678.

Fox, P. T., & Friston, K. J. (2012). Distributed processing: Distributed functions? *NeuroImage, 61*(2), 407–426.

Fried, I., MacDonald, K. A., & Wilson, C. L. (1997). Single neuron activity in human hippocampus and amygdala during recognition of faces and objects. *Neuron, 18*(5), 753–765.

Galambos, R. (1996). Autobiography. In L. Squires (Ed.), *The History of Neuroscience in Autobiography* (Vol. 1, pp. 178–221). Washington, DC: Society for Neuroscience.

Galambos, R. (2007). Reflections on the conceptual origins of neuron-glia interactions. *Neuron Glia Biology, 3*, 89–91.

Gall, F. J., & Spurzheim, J. C. (1808). *Recherches sur le systeme nerveux en general et sur celui du cerveau en particulier.* Paris: Memoirs de Academie de Sciences.

Gallese, V., Fadiga, L., Fogassi, L., & Rizzolatti, G. (1996). Action recognition in the premotor cortex. *Brain, 119*, 593–609.

Gerard, R. W., & Forbes, A. (1928). "Fatigue" of the flexion reflex. *American Journal of Physiology, 86*(1), 186–205.

Gerstein, G. L., & Clark, W. A. (1964). Simultaneous studies of firing patterns in several neurons. *Science, 143*(361), 1325–1357.

Gill, S. (1958). Parallel programming. *Computer Journal, 1*(1), 2–10.

Gingrich, K. J., Baxter, D. A., & Byrne, J. H. (1988). Mathematical-model of cellular mechanisms contributing to presynaptic facilitation. *Brain Research Bulletin, 21*(3), 513–520.

Gingrich, K. J., & Byrne, J. H. (1985). Simulation of synaptic depression, posttetanic potentiation, and presynaptic facilitation of synaptic potentials from sensory neurons mediating gill-withdrawal reflex in Aplysia. *Journal of Neurophysiology, 53*(3), 652–669.

Goertzel, B., Lian, R. T., Arel, I., de Garis, H., & Chen, S. (2010). A world survey of artificial brain projects, Part II: Biologically inspired cognitive architectures. *Neurocomputing, 74*(1–3), 30–49.

Goldstein, H. H., & Goldstein, A. (1946). The ENIAC. *M.T.A.C., 2*, 97–110.

Golgi, C. (1906). *The neuron doctrine—theory and facts.* Paper presented at the 1906 Nobel Foundation Awards Ceremony. Stockholm, Sweden.

Gonzalez Castillo, J., Saad, Z. S., Handwerker, D. A., Inati, S. J., Brenowitz, N., & Bandettini, P. A. (2012). Whole-brain, time-locked activation with simple tasks revealed with massive averaging and model-free analysis. *Proceedings of the National Academy of Sciences of the United States of America, 109*, 5487–5492.

Goodman, M. B., Hall, D. H., Avery, L., & Lockery, S. R. (1998). Active currents regulate sensitivity and dynamic range in C. elegans neurons. *Neuron, 20*(4), 763–772.

Graham, J., & Gerard, R. W. (1946). *Journal of Cellular and Comparative Physiology, 28*, 99.

Gross, C. G. (2002). Genealogy of the "grandmother cell." *Neuroscientist, 8*(5), 512–518.

Gross, G. W., Rieske, E., Kreutzberg, G. W., & Meyer, A. (1977). New fixed-array multi-microelectrode system designed for long-term monitoring of extracellular single unit neuronal-activity in vitro. *Neuroscience Letters, 6*(2–3), 101–105.

Gross, G. W., Williams, A. N., & Lucas, J. H. (1982). Recording of spontaneous activity with photoetched microelectrode surfaces from mouse spinal neurons in culture. *Journal of Neuroscience Methods, 5*(1–2), 13–22.

Grossberg, S., & Mingolla, E. (1985). Neural dynamics of perceptual grouping: Textures, boundaries, and emergent segmentations. *Perception & Psychophysics, 38*(2), 141–171.

Grundfest, H., Sengstaken, R. W., Oettinger, W. H., & Gurry, R. W. (1950). Stainless steel micro-needle electrodes made by electrolytic pointing. *Review of Scientific Instruments, 21*(4), 360–361.

Ha, H. I., Hendricks, M., Shen, Y., Gabel, C. V., Fang-Yen, C., Qin, Y. Q., . . . Zhang, Y. (2010). Functional organization of a neural network for aversive olfactory learning in Caenorhabditis elegans. *Neuron, 68*(6), 1173–1186.

Ham, F. M., & Kostanic, I. (2000). *Principles of Neurocomputing in Science and Engineering*. New York: McGraw-Hill.

Harary, F. (1969). *Graph Theory*. Reading, MA: Addison-Wesley Pub. Co.

Harris, J. D. (1943). Habituatory response decrement in the intact organism. *Psychological Bulletin, 40*(6), 385–422.

Hartline, H. K. (1938). The response of single optic nerve fibers of the vertebrate eye to illumination of the retina. *American Journal of Physiology, 121*(2), 400–415.

Hartline, H. K. (1940). The effects of spatial summation in the retina on the excitation of the fibers of the optic nerve. *American Journal of Physiology, 130*(4), 700–711.

Hartline, H. K. (1949). Inhibition of activity of visual receptors by illuminating nearby retinal areas in the Limulus eye. *Federation Proceedings, 8*(1), 69–69.

Hartline, H. K., & Graham, C. H. (1932). Nerve impulses from single receptors in the eye. *Journal of Cellular and Comparative Physiology, 1*(2), 277–295.

Hartline, H. K., & Ratliff, F. (1957). Inhibitory interaction of receptor units in the eye of Limulus. *Journal of General Physiology, 40*, 357–376.

Hartline, H. K., & Ratliff, F. (1958). Spatial summation of inhibitory influences in the eye of Limulus, and the mutual interaction of receptor units. *Journal of General Physiology, 41*, 1049–1066.

Hartline, H. K., Wagner, H. G., & Ratliff, F. (1956). Inhibition in the eye of Limulus. *Journal of General Physiology, 39*(5), 651–673.

Hebb, D. O. (1949). *The Organization of Behavior: A Neuropsychological Theory*. New York: Wiley.

Heer, F., Franks, W., Blau, A., Taschini, S., Ziegler, C., Hierlemann, A., & Baltes, H. (2004). CMOS microelectrode array for the monitoring of electrogenic cells. *Biosensors & Bioelectronics, 20*(2), 358–366.

Hell, S. W., & Wichmann, J. (1994). Breaking the diffraction resolution limit by stimulated-emission—Stimulated-emission-depletion fluorescence microscopy. *Optics Letters, 19*(11), 780–782.

Henson, R. (2006). Forward inference using functional neuroimaging: Dissociations versus associations. *TRENDS in Cognitive Sciences, 10*(2), 64–69.

Hertz, J., Krogh, A., & Palmer, R. (1991). *Introduction to the Theory of Neural Computation.* Redwood City, CA: Addison-Wesley.

Hickok, G. (2009). Eight problems for the mirror neuron theory of action understanding in monkeys and humans. *Journal of Cognitive Neuroscience, 21*(7), 1229–1243.

Hilgetag, C. C., O'Neill, M. A., & Young, M. P. (1996). Indeterminate organization of the visual system—Rejoinder. *Science, 271*(5250), 777–777.

Hinton, G. E., & Sejnowski, T. J. (1986). Learning and relearning in Boltzmann machines. In D. E. Rumelhart & J. L. McClelland (Eds.), *Parallel Distributed Processing. Explorations in the Microstructure of Cognition. Vol. 1: Foundations.* Cambridge, MA: MIT Press.

Hodgkin, A. L., & Huxley, A. F. (1952). Currents carried by sodium and potassium ions through the membrane of the giant axon of Loligo. *Journal of Physiology, 116*(4), 449–472.

Hodgkin, A. L., Huxley, A. F., & Katz, B. (1952). Measurement of current-voltage relations in the membrane of the giant axon of Loligo. *Journal of Physiology, 116*(4), 424–448.

Hodgkin, A. L., & Katz, B. (1949). The effect of sodium ions on the electrical activity of giant axon of the squid. *Journal of Physiology, 108*(1), 37–77.

Hoogerwerf, A. C., & Wise, K. D. (1994). A 3-dimensional microelectrode array for chronic neural recording. *IEEE Transactions on Biomedical Engineering, 41*(12), 1136–1146.

Hopfield, J. J. (1982). Neural networks and physical systems with emergent collective computational abilities. *Proceedings of the National Academy of Sciences of the United States of America-Biological Sciences, 79*(8), 2554–2558.

Horvath, J. C., Carter, O., & Forte, J. D. (2014). Transcranial direct current stimulation: Five important issues we aren't discussing (but probably should be). [Review]. *Frontiers in Systems Neuroscience, 8*, 2.

Horvath, J. C., Forte, J. D., & Carter, O. (2015a). Quantitative review finds no evidence of cognitive effects in healthy populations from single-session transcranial; direct current stimulation (tDCS). *Brain Stimulation, 8*(3), 535–550.

Horvath, J. C., Forte, J. D., & Carter, O. (2015b). Evidence that transcranial direct current stimulation (tDCS) generates little-to-no reliable neurophysiologic effect beyond MEP amplitude modulation in healthy human subjects: A systematic review. *Neuropsychologia, 66*, 213–236.

Horwitz, B. (2003). The elusive concept of brain connectivity. *NeuroImage, 19*(2), 466–470.

Hubel, D. H. (1957). Tungsten microelectrode for recording from single units. *Science, 125*(3247), 549–550.

Hubel, D. H., & Wiesel, T. N. (1959). Receptive fields of single neurones in the cat's striate cortex. *The Journal of Physiology, 148*, 574–571.

Hubel, D. H., & Wiesel, T. N. (1965). Receptive fields and functional architecture in 2 nonstriate visual areas (18 and 19) of cat. *Journal of Neurophysiology, 28*(2), 229–289.

Hubel, D. H., & Wiesel, T. N. (1968). Receptive fields and functional architecture of monkey striate cortex. *Journal of Physiology-London, 195*(1), 215–243.

Izhikevich, E. M. (2003). Simple model of spiking neurons. *IEEE Transactions on Neural Networks, 14*(6), 1569–1572.

Izhikevich, E. M., & Edelman, G. M. (2008). Large-scale model of the mammalian thalamocortical system. *PNAS, 105*, 3593–3598.

James, W., & Rouben Mamoulian Collection (Library of Congress). (1890). *The Principles of Psychology*. New York: H. Holt & Co.

Jenison, R. L., Rangel, A., Oya, H., Kawasaki, H., & Howard, M. A. (2011). Value encoding in single neurons in the human amygdala during decision making. *Journal of Neuroscience, 31*(1), 331–338.

Johnson, M. H. (2011). Interactive specialization: A domain-general framework for human functional brain development? *Developmental Cognitive Science 1*(1):7–21. doi:10.1016/j.dcn.2010.07.003. Epub 2010 Jul 30.

Johnstone, A. F. M., Gross, G. W., Weiss, D. G., Schroeder, O. H. U., Gramowski, A., & Shafer, T. J. (2010). Microelectrode arrays: A physiologically based neurotoxicity testing platform for the 21st century. *Neurotoxicology, 31*(4), 331–350.

Jones, D. K., Simmons, A., Williams, S. C. R., & Horsfield, M. A. (1999). Non-invasive assessment of axonal fiber connectivity in the human brain via diffusion tensor MRI. *Magnetic Resonance in Medicine, 42*(1), 37–41.

Jones, K. E., Campbell, P. K., & Normann, R. A. (1992). A glass silicon composite intracortical electrode array. *Annals of Biomedical Engineering, 20*(4), 423–437.

Jung, R., & Spillmann, L. (1970). Receptive-field estimation and perceptual integration in human vision. In F. A. Young & D. B. Lindsley (Eds.), *Early Experience and Visual Information Processing Perceptual and Reading Disorders* (pp. 181–197). Washington, DC: National Academy of Sciences.

Kandel, E. R. (1991). Cellular mechanisms of learning and the biological basis of individuality. In E. R. Kandel & J. H. Schwartz (Eds.), *Principles of Neural Science (3rd Edition)* (pp. 1009–1032). New York: Elsevier.

Kandel, E. R., & Tauc, L. (1964). Mechanism of prolonged heterosynaptic facilitation. *Nature, 202*(492), 145–147.

Karp, R. M. (1986). Combinatorics, complexity, and randomness. *Communications of the Acm, 29*(2), 98–109.

Karr, J. & Herman, D. (2004). Using a MEA to analyze a dissociated hipposcampal neural network. http://web.mit.edu/9.29/www/neville_jen/jkarr/mea/

Keefer, E. W., Botterman, B. R., Romero, M. I., Rossi, A. F., & Gross, G. W. (2008). Carbon nanotube coating improves neuronal recordings. *Nature Nanotechnology, 3*(7), 434–439.

Kennard, M. A. (1955). Effect of bilateral ablation of cingulate area on behavior. *Journal of Neurophysiology, 18,* 159–169.

Kibler, A. B., Jamieson, B. G., & Durand, D. M. (2012). A high aspect ratio microelectrode array for mapping neural activity in vitro. *Journal of Neuroscience Methods, 204*(2), 296–305.

Killeen, P. R. (2013). The structure of scientific evolution. *Behavior Analyst, 36*(2), 325–344.

Ko, K. I. (1984). On the computational complexity of ordinary differential equations. *Information and Control, 58*, 157–194.

Koch, C. (2012). Modular biological complexity. *Science, 337*, 531–532.

Kohl, J., & Jefferis, G. S. X. E. (2010). Neuroranatomy. *Current Biology, 21*, R19–R20.

Kohonen, T. (1977). *Associative Memory: A System-Theoretical Approach*. Berlin: Springer Verlag.

Konorski, J. (1967). *Integrative Activity of the Brain: An Interdisciplinary Approach*. Chicago: University of Chicago Press.

Kreiman, G., Koch, C., & Fried, I. (2000). Category-specific visual responses of single neurons in the human medial temporal lobe. *Nature Neuroscience, 3*(9), 946–953.

Kruger, J., & Bach, M. (1981). Simultaneous recording with 30 microelectrodes in monkey visual-cortex. *Experimental Brain Research, 41*(2), 191–194.

Kupfermann, I., Castellucci, V., Pinsker, H., & Kandel, E. (1970). Neuronal correlates of habituation and dishabituation of gill-withdrawal reflex in Aplysia. *Science, 167*(3926), 1743–1745.

Kwong, K. K. (2012). Record of a single fMRI experiment in May of 1991. *NeuroImage, 62*(2), 610–612.

Lazer, D. (2015). The rise of the social algorithm. *Science, 348*, 1090–1091.

Leopold, D. A., Bondar, I. V., & Giese, M. A. (2006). Norm-based face encoding by single neurons in the monkey inferotemporal cortex. *Nature, 442*(7102), 572–575.

Lettvin, J. Y., Maturana, H. R., Mcculloch, W. S., & Pitts, W. H. (1959). What the frogs eye tells the frogs brain. *Proceedings of the Institute of Radio Engineers, 47*(11), 1940–1951.

Lewicki, M. S. (1998). A review of methods for spike sorting: The detection and classification of neural action potentials. *Network-Computation in Neural Systems, 9*(4), R53–R78.

Lichtman, J. W. & Denk, W. (2011). The big and the small: challenges of imaging the brain's circuits. *Science, 334*, 618–623.

Lindquist, K. A., Wager, T. D., Kober, H., Bliss-Moreau, E., & Barrett, L. F. (2012). The brain basis of emotion: A meta-analytic review. *Behavioral and Brain Sciences, 35*(3), 121–143.

Ling, G., & Gerard, R. W. (1949). The normal membrane potential of frog sartorius fibers. *Journal of Cellular and Comparative Physiology, 34*, 383–396.

Litke, A. M., Bezayiff, N., Chichilnisky, E. J., Cunningham, W., Dabrowski, W., Grillo, A. A., . . . Sher, A. (2004). What does the eye tell the brain? Development of a system for the large-scale recording of retinal output activity. *IEEE Transactions on Nuclear Science, 51*(4), 1434–1440.

Lockery, S. R., & Goodman, M. B. (1998). Tight-seal whole-cell patch clamping of Caenorhabditis elegans neurons. *Ion Channels, Pt B, 293*, 201–217.

Logothetis, N. K., Pauls, J., Augath, M., Trinath, T., & Oeltermann, A. (2001). Neurophysiological investigation of the basis of the fMRI signal. [Research Support, Non-U.S. Gov't]. *Nature, 412*(6843), 150–157.

Lorente de No, R. (1938). Synaptic stimulation of motoneurons as a local process. *Journal of Neurophysiology, 1*, 195–206.

Lumer, E. D., Edelman, G. M., & Tononi, G. (1997a). Neural dynamics in a model of the thalamocortical system. 1. Layers, loops and the emergence of fast synchronous rhythms. *Cerebral Cortex*, 7(3), 207–227.

Lumer, E. D., Edelman, G. M., & Tononi, G. (1997b). Neural dynamics in a model of the thalamocortical system. 2. The role of neural synchrony tested through perturbations of spike timing. *Cerebral Cortex*, 7(3), 228–236.

Luo, Y. H., & da Cruz, L. (2014). A review and update on the current status of retinal prostheses (bionic eye). *British Medical Bulletin, 109*, 31–44.

Maccione, A., Garofalo, M., Nieus, T., Tedesco, M., Berdondini, L., & Martinoia, S. (2012). Multiscale functional connectivity estimation on low-density neuronal cultures recorded by high-density CMOS Micro Electrode Arrays. *Journal of Neuroscience Methods, 207*(2), 161–171.

MacCorquodale, K., & Meehl, P. E. (1948). On a distinction between hypothetical constructs and intervening variables. *Psychological Review, 55*(2), 95–107.

MacNichol, E. F., Jr. (1956). Visual receptors as biological transducers. In R. G. Grenell & L. J. Mullins (Eds.), *Molecular Structure and Functional Activity of Nerve Cells* (Publication 1, 34). Washington, D.C.: American Institute of Biological Sciences.

Margolis, D. J., Gartland, A. J., Singer, J. H., & Detwiler, P. B. (2014). Network oscillations drive correlated spiking of ON and OFF ganglion cells in the rd1 mouse model of retinal degeneration. *PLoS One, 9*(1), e86253. doi:10.1371/journal.pone.0086253

Markram, H. (2006). The Blue Brain Project. *Nature Reviews Neuroscience, 7*(2), 153–160.

Martin, S. J., Grimwood, P. D., & Morris, R. G. (2000). Synaptic plasticity and memory: An evolution of the hypothesis. *Annual Review Neuroscience, 23*, 649–711.

McClelland, J. L., & Rumelhart, D. E. (1988). A simulation-based tutorial system for exploring parallel distributed-processing. *Behavior Research Methods Instruments & Computers, 20*(2), 263–275.

McClelland, J. L., Rumelhart, D. E., & The PDP Research Group (1988). *Parallel Distributed Processing: Explorations in the Microstructure of Cognition. Vol 2 Psychological and Biological Models.* Cambridge, MA: MIT Press.

McCulloch, W. S. (1944). The functional organization of the cerebral cortex. *Physiological Reviews, 24*(3), 390–407.

McCulloch, W. S., & Pitts, W. (1943). A logical calculus of the ideas immanent in nervous activity. *Bulletin of Mathematical Biophysics, 5*, 115–133.

Merolla, P. A., Arthur, J. V., Alvarez-Icaza, R., Cassidy, A. S., Sawada, J., Akopyan, F., . . . Modha, D. S. (2014). Artificial brains: A million spiking-neuron integrated circuit with a scalable communication network and interface. *Science, 345*(6197), 668–673.

Meyer, A. R. (1974). *The inherent computational complexity of theories of ordered sets.* Paper presented at the Proceedings of the International Congress of Mathematicians, Vancouver, Canada.

Minsky, M. L., & Papert, S. (1969). *Perceptrons: An Introduction to Computational Geometry.* Cambridge, MA: MIT Press.

Moore, E. F. (1956). *Gedanken—experiments on sequential machines.* In C. E. Shannon & J. McCarthy (Eds.), *Automata Studies* (pp. 129–153). Princeton, NJ: Princeton University Press.

Moore, G. E. (1965). Cramming more components onto integrated circuits. *Electronics, 38*(8), 114–117.

Moxon, K. A., Hallman, S., Sundarakrishnon, A., Wheatley, M., Nissanov, J., & Barbee, K. A. (2009). Long-term recordings of multiple single-neurons for clinical applications: The emerging role of the bioactive microelectrode. *Materials, 2*, 1762–1794.

Mukamel, D., Ruffo, S., & Schreiber, N. (2005). Breaking of ergodicity and long relaxation times in systems with long-range interactions. *Physical Review Letters, 95*(24), 240604.

Murphy, T. H., Blatter, L. A., Wier, W. G., & Baraban, J. M. (1992). Spontaneous synchronous synaptic calcium transients in cultured cortical-neurons. *Journal of Neuroscience, 12*(12), 4834–4845.

Nabavi, S., Fox, R., Perolux, C. D., Lin, J. Y., Tsien, R. Y., & Malinow, R. (2014). Engineering a memory with LTD and LTP. *Nature, 511*, 348–352.

Neri, P., & Levi, D. M. (2006). Receptive versus perceptive fields from the reverse-correlation viewpoint. *Vision Res, 46*(16), 2465–2474.

Nicolelis, M. A. L., & Chapin, J. K. (1994). Spatiotemporal structure of somatosensory responses of many-neuron ensembles in the rat ventral posterior medial nucleus of the thalamus. *Journal of Neuroscience, 14*(6), 3511–3532.

Nordhausen, C. T., Maynard, E. M., & Normann, R. A. (1996). Single unit recording capabilities of a 100 microelectrode array. *Brain Research, 726*(1–2), 129–140.

Ogawa, S., Lee, T. M., Kay, A. R., & Tank, D. W. (1990). Brain magnetic-resonance-imaging with contrast dependent on blood oxygenation. *Proceedings of the National Academy of Sciences of the United States of America, 87*(24), 9868–9872.

Olsen, S. R., & Wilson, R. I. (2008). Cracking neural circuits in a tiny brain: New approaches for understanding the neural circuitry of Drosophila. *Trends in Neurosciences, 31*(10), 512–520.

Oosterwijk, S., Lindquist, K. A., Anderson, E., Dautoff, R., Moriguchi, Y., & Bennett, L. F. (2012). States of mind: Emotions, body feelings, and thoughts share distributed neural networks. *NeuroImage 62*, 2110–2128.

Palmer, S. E. (1999). *Vision science: Photons to phenomenology*. Cambridge, MA: MIT.

Papadimitriou, C. H., & Tsitsiklis, J. (1986). Intractable problems in control-theory. *Siam Journal on Control and Optimization, 24*(4), 639–654.

Perrett, D. I., Rolls, E. T., & Caan, W. (1982). Visual neurones responsive to faces in the monkey temporal cortex. *Experimental Brain Research, 47*(3), 329–342.

Pickel, V. M., & Segal, M. (Eds.). (2013). *The Synapse: Structure and Function*. Oxford, UK: Academic Press.

Pillow, J. W., Shlens, J., Chichilnisky, E. J., & Simoncelli, E. P. (2013). A model-based spike sorting algorithm for removing correlation artifacts in multi-neuron recordings. *PLoS One, 8*(5), e62123. doi:10.1371/journal.pone.0062123

Pine, J. (1980). Recording action-potentials from cultured neurons with extracellular micro-circuit electrodes. *Journal of Neuroscience Methods, 2*(1), 19–31.

Pinsker, H., Kupfermann, I., Castellucci, V., & Kandel, E. (1970). Habituation and dishabituation of the gill withdrawal reflex in Aplysia. *Science, 167*, 1740–1732.

Pitts, W., & McCulloch, W. S. (1947). How we know universals: The perception of auditory and visual forms. *Bulletin of Mathematical Biophysics, 9*, 127–147.

Place, U. T. (1956). Is consciousness a brain process? *British Journal of Psychology, 47*(1), 44–50.

Poldrack, R. A. (2006). Can cognitive processes be inferred from neuroimaging data? *Trends in Cognitive Sciences, 10*(2), 59–63.

Poldrack, R. A. (2010). Mapping mental function to brain structure: How can cognitive neuroimaging succeed? *Perspectives on Psychological Science, 5*(6), 753–761.

Posner, M. I., Petersen, S. E., Fox, P. T., & Raichle, M. E. (1998). Localization of cognitive operations in the human brain. *Science, 240*, 1627–1631.

Quine, W. V. O. (1951). Main trends in recent philosophy. *Philosophical Review, 60*, 20–43.

Quiroga, R. Q. (2012). Concept cells: The building blocks of declarative memory functions. *Nature Reviews Neuroscience, 13*(8), 587–597.

Quiroga, R. Q., Nadasdy, Z., & Ben-Shaul, Y. (2004). Unsupervised spike detection and sorting with wavelets and superparamagnetic clustering. *Neural Computation, 16*(8), 1661–1687.

Quiroga, R. Q., Reddy, L., Kreiman, G., Koch, C., & Fried, I. (2005). Invariant visual representation by single neurons in the human brain. *Nature, 435*(7045), 1102–1107.

Rabi, I. I., Zacharias, J. R., Millman, S., & Kusch, P. (1937). A new method for measuring nuclear magnetic moments. *Physical Review, 53,* 318.

Rabi, I. I., Zacharias, J. R., Millman, S., & Kusch, P. (1939). The molecular beam resonance method for measuring nuclear magnetic moments. The magnetic moments of 3LI6, 3LI7, and 9F19. *Physical Review, 55,* 526–535.

Raizen, D. M., & Avery, L. (1994). Electrical-activity and behavior in the pharynx of C. elegans. *Neuron, 12*(3), 483–495.

Rakover, S. S. (2011). A plea for methodological dualism and a multi-explanation approach in psychology. *Behavior and Philosophy, 39,* 17–43.

Ratliff, F., Hartline, H. K., & Miller, W. H. (1963). Spatial and temporal aspects of retinal inhibitory interaction. *Journal of the Optical Society of America, 53,* 110–120.

Richmond, J. E. (2006). Electrophysiological recordings from the neuromuscular junction of *C. elegans.* In The *C. elegans* Research Community (Ed.), *WormBook,* 1–8.

Rosenblatt, F. (1958). The perceptron—A probabilistic model for information-storage and organization in the brain. *Psychological Review, 65*(6), 386–408.

Roskies, A. L. (2010). How does neuroscience affect our conception of volition? *Annual Review of Neuroscience, 33,* 190–130.

Rousche, P. J., & Normann, R. A. (1998). Chronic recording capability of the Utah intracortical electrode array in cat sensory cortex. *Journal of Neuroscience Methods, 82*(1), 1–15.

Rumelhart, D. E., Hinton, G. E., & Williams, R. J. (1986). Learning representations by back-propagating errors. *Nature, 323*(6088), 533–536.

Rumelhart, D. E., McClelland, J. L., & The PDP Research Group (1988). *Parallel Distributed Processing. Explorations in the Microstructure of Cognition. Vol 1: Foundations.* Cambridge, MA: MIT Press.

Schwarz, D. A., Lebedev, M. A., Hanson, T. L., Dimitrov, D. F., Lehew, G., Meloy, J., . . . Nicolelis, M. A. (2014). Chronic, wireless recordings of large-scale brain activity in freely moving rhesus monkeys. *Nature Methods, 11,* 670–676.

Searle, J. R. (1990). *The Mystery of Consciousness.* New York: New York Review of Books.

Searle, J. R. (1997, March 6). *Consciousness and the Philosophers.* New York: New York Review of Books.

Sejnowski, T. J., Kienker, P. K., & Hinton, G. E. (1986). Learning symmetry groups with hidden units—Beyond the perceptron. *Physica D-Nonlinear Phenomena, 22*(1–3), 260–275.

Seker, E., Berdichevsky, Y., Begley, M. R., Reed, M. L., Staley, K. J., & Yarmush, M. L. (2010). The fabrication of low-impedance nanoporous gold multiple-electrode arrays for neural electrophysiology studies. *Nanotechnology, 21*(12). doi:10.1088/0957-4484/21/12/125504

Selfridge, O. G. (1958). *Pandemonium: A paradigm for learning.* Paper presented at the Proceeding of a Symposium the Mechanization of Thought Processes, London.

Service, R. F. (2014). The brain chip. *Science, 345*(6197), 614–616.

Sevush, S. (2006). Single-neuron theory of consciousness. *Journal of Theoretical Biology, 238*(3), 704–725.

Sherrington, C. S. (1940). *Man on His Nature.* Cambridge, UK: The University Press.

Shlens, J., Field, G. D., Gauthier, J. L., Greschner, M., Sher, A., Litke, A. M., & Chichilnisky, E. J. (2009). The structure of large-scale synchronized firing in primate retina. *Journal of Neuroscience, 29*(15), 5022–5031.

Simmons, J. P., Nelson, L. D., & Simonsohn, U. (2011). False-positive psychology: Undisclosed flexibility in data collection and analysis allows presenting anything as significant. *Psychological Science, 22*(11), 1359–1366.

Skinner, B. F. (1950). Are theories of learning necessary? *Psychological Review, 57*, 193–216.

Spillmann, L. (2014). Receptive fields of visual neurons: The early years. [Research Support, Non-U.S. Gov't]. *Perception, 43*(11), 1145–1176.

Spillmann, L., Dresp-Langley, B., & Tseng, C.-H. (2015). Beyond the classical receptive field: The effect of contextural stimuli. *Journal of Vision, 15*(9), 7, 1–23.

Spillmann, L., & Werner, J. S. (Eds.). (1990). *Visual Perception: The Neurophysiological Foundations*. San Diego: Academic Press.

Spira, M. E., & Hai, A. (2013). Multi-electrode array technologies for neuroscience and cardiology. *Nature Nanotechnology, 8*(2), 83–94.

Sporns, O. (2011). *Networks of the Brain*. Cambridge, MA: MIT Press.

Sporns, O., Tononi, G., & Kotter, R. (2005). The human connectome: A structural description of the human brain. *PLoS Computational Biology, 1*(4), 245–251.

Steidl, E. M., Neveu, E., Bertrand, D., & Buisson, B. (2006). The adult rat hippocampal slice revisited with multi-electrode arrays. *Brain Research, 1096*, 70–84.

Stevenson, I. H., Rebesco, J. M., Hatsopoulos, N. G., Haga, Z., Miller, L. E., & Kording, K. P. (2009). Bayesian inference of functional connectivity and network structure from spikes. *IEEE Transactions on Neural Systems and Rehabilitation Engineering, 17*(3), 203–213.

Stockmeyer L. J., & Chandra, A. K. (1979). Provably difficult combinatorial games, *SIAM Journal on Computing, 8*(2), 151–174.

Takeuchi, T., Duszkiewicz, A. J., & Morris, R. G. (2013). The synaptic plasticity and memory hypothesis: Encoding, storage and persistence. [Research Support, Non-U.S. Gov't]. *Philosophical Transactions of the Royal Society of London Series B-Biological Sciences, 369*(1633), 20130288.

Tanaka, K. (1993). Neuronal mechanisms of object recognition. *Science, 262*(5134), 685–688.

Thomas, C. A., Jr., Springer, P. A., Loeb, G. E., Berwald-Netter, Y., & Okun, L. M. (1972). A miniature microelectrode array to monitor the bioelectric activity of cultured cells. *Experimental Cell Research, 74*(1), 61–66.

Thompson, R. F., & Spencer, W. A. (1966). Habituation—A model phenomenon for study of neuronal substrates of behavior. *Psychological Review, 73*(1), 16–43.

Thyreau, B., Schwartz, Y., Thirionm, B., Frouin, V., Loth, E., Vollstädt-Klein, S., . . . IMAGEN Consortium. (2012). Very large fMRI study using the IMAGEN database: Sensitivity-specificity and population effect modeling in relation to the underlying anatomy. *NeuroImage, 61*(1), 295–303.

Turella, L., Pierno, A. C., Tubaldi, F., & Castiello, U. (2009). Mirror neurons in humans: Consisting or confounding evidence? *Brain and Language, 108*(1), 10–21.

Turing, A. (1950). Computing machinery and intelligence. *Mind, 59*, 433–460.

Turing, A. (2004). Lecture on the automatic computing engine. In B. J. Copeland (Ed.), *The Essential Turing* (pp. 362–394). Oxford: Oxford University Press.

Turkeltaub, P. E., Eden, G. F., Jones, K. M., & Zeffiro, T. A. (2002). Meta-analysis of the functional neuroanatomy of single-word reading: Method and validation. *NeuroImage, 16*(3), 765–780.

Underwood, E. (2014). Brain Project meets physics. *Science, 344*, 954–955.

Uttal, W. R. (1959). A comparison of neural and psychophysical responses in the somesthetic system. *Journal of Comparative and Physiological Psychology, 52*(4), 485–490.

Uttal, W. R. (1973). *The Psychobiology of Sensory Coding*. New York: Harper & Row.

Uttal, W. R. (1975). *An Autocorrelation Theory of Form Detection*. Hillsdale, NJ and New York: L. Erlbaum Associates; Distributed by Halsted Press.

Uttal, W. R. (1978). *The Psychobiology of Mind*. Hillsdale, NJ: Erlbaum.

Uttal, W. R. (1990). On some 2-way barriers between models and mechanisms. *Perception & Psychophysics, 48*(2), 188–203.

Uttal, W. R. (2001). *The New Phrenology: The Limits of Localizing Cognitive Processes in the Brain*. Cambridge, MA: MIT Press.

Uttal, W. R. (2004). *Dualism: The Original Sin of Cognitivism*. Mahwah, NJ: Erlbaum.

Uttal, W. R. (2005). *Neural Theories of Mind*. Mahwah, NJ: Erlbaum.

Uttal, W. R. (2009). *Distributed Neural Systems: Beyond the New Phrenology*. Cornwall-on-Hudson, NY: Sloan Pub.

Uttal, W. R. (2011). *Mind and Brain: A Critical Appraisal of Cognitive Neuroscience*. Cambridge, MA: MIT Press.

Uttal, W. R. (2013). *Reliability in Cognitive Neuroscience: A Meta-Meta-Analysis*. Cambridge, MA: MIT Press.

Uttal, W. R. (2016). *Macroneural Theories in Cognitive Neuroscience*. New York: Psychology Press.

Uttal, W. R., & Krissoff, M. (1966). Effect of stimulus pattern on temporal acuity in somatosensory system. *Journal of Experimental Psychology, 71*(6), 878–883.

Vacariu, G. (2012). *Cognitive Neuroscience Versus Epistemologically Different Worlds*. Bucharest, Romania: University of Bucharest Press.

Vacariu, G., & Vacariu, M. (2013). Troubles with cognitive neuroscience. *Philosophia Scientiae, 17*, 151–170.

Valbo, A. B., & Hagbarth, K. E. (1968). Activity from skin mechanoreceptors recorded percutaneously in a wake human subjects. *Experimental Neurology, 21*, 270–289.

Van Essen, D. C., Anderson, C. H., & Felleman, D. J. (1992). Information processing in the primate visual system: An integrated systems perspective. *Science, 255*, 419–423.

Van Orden, G. C., & Papp, K. R. (1997). Functional neuroimages fail to discover pieces of the mind in the parts of the brain. *Philosophy of Science Proceedings, 64*, 585–594.

Vimal, R. L. P. (2009). Meanings Attributed to the Term "Consciousness": An overview. *Journal of Consciousness Studies, 16*(5), 9–27.

Vogelstein, J. T., Park, Y., Ohyama, T., Kerr, R. A., Truman, J. W., Priebe, C. E., & Zlatic, M. (2014). Discovery of brainwide neural-behavioral maps via multiscale unsupervised structure learning. [Research Support, Non-U.S. Gov't, U.S. Gov't, Non-P.H.S.]. *Science, 344*(6182), 386–392.

von Neumann, J. (1945). *First Draft of a Report on the EDVAC*. Philadelphia: University of Pennsylvania.

von Philipsborn, A. C., Liu, T., Yu, J. Y., Masser, C., Bidaye, S. S., & Dickson, B. J. (2011). Neuronal control of Drosophila courtship song. [Comparative Study Research Support, Non-U.S. Gov't]. *Neuron, 69*(3), 509–522.

Wagenaar, D. A., Madhavan, R., Pine, J., & Potter, S. M. (2005). Controlling bursting in cortical cultures with closed-loop multi-electrode stimulation. *Journal of Neuroscience, 25*(3), 680–688.

Wagenaar, D. A., Pine, J., & Potter, S. M. (2006). Searching for plasticity in dissociated cortical cultures on multi-electrode arrays. [Research Support, N.I.H., Extramural Research Support, Non-U.S. Gov't]. *Journal of Negative Results Biomedicine, 5*, 16.

Ward, A. A., & Thomas, L. B. (1955). The electrical activity of single units in the cerebral cortex of man. *Electroencephalography and Clinical Neurophysiology,* 7(1), 135–136.

Wark, H. A., Sharma, R., Mathews, K. S., Fernandez, E., Yoo, J., Christensen, B., Tresco, P. . . . Tathireddy, P. (2013). A new high-density (25 electrodes/mm^2) penetrating microelectrode array for recording and stimulating sub-millimeter neuroanatomical structures. *Journal of Neural Engineering, 10*(4):045003. doi:10.1088/1741-2560/10/4/045003.

Watson, J. D., & Crick, F. H. C. (1953). Molecular structure of nucleic acids—A structure for deoxyribose nucleic acid. *Nature, 171*(4356), 737–738.

Westheimer, G. (1990). Relating neural mechanisms to visual perception. In L. Spillman & J. S. Werner (Eds.), *Visual Perception: The Neurophysiological Foundations.* San Diego: Academic Press.

White, J. G., Southgate, E., Thomson, J. N., & Brenner, S. (1986). The structure of the nervous-system of the nematode C. elegans. *Philosophical Transactions of the Royal Society of London Series B-Biological Sciences, 314*(1165), 1–340.

Wild, J., Prekopcsak, Z., Sieger, T., Novak, D., & Jech, R. (2012). Performance comparison of extracellular spike sorting algorithms for single-channel recordings. *Journal of Neuroscience Methods, 203*(2), 369–376.

Wilson-Sanders, S. E. (2011). Invertebrate models for biomedical research, testing, and education. *ILAR Journal, 52*(2), 126–152.

Yoshimura, Y., Dantzker, J. L. M., & Callaway, E. M. (2005). Excitatory cortical neurons form fine-scale functional networks. *Nature, 433*(7028), 868–873.

Young, J. Z. (1936). The giant nerve fibres and epistellar body of cephalopods. *Quarterly Journal of Microscopical Science, 78*, 367–386.

Young, M. P., & Yamane, S. (1992). Sparse population coding of faces in the inferotemporal cortex. *Science, 256*(5061), 1327–1331.

Yule, G. U. (1926). Why do we sometimes get nonsense-correlations between time-series?—A study in sampling and the nature of time-series. *Journal of the Royal Statistical Society, 89*, 1–69.

Zecevic, D., Wu, J. Y., Cohen, L. B., London, J. A., Hopp, H. P., & Falk, C. X. (1989). Hundreds of neurons in the *Aplysia* abdominal-ganglion are active during the gill-withdrawal reflex. *Journal of Neuroscience, 9*(10), 3681–3689.

Zeng, F. G., Rebscher, S., Harrison, W., Sun, X., & Feng, H. (2008). Cochlear implants: System design, integration, and evaluation. [Research Support, N.I.H., Extramural Research Support, Non-U.S. Gov't Review]. *IEEE Reviews in Biomedical Engineering, 1*, 115–142.

Zhang, Y. L., Liu, R. Y., Heberton, G. A., Smolen, P., Baxter, D. A., Cleary, L. J., & Byrne, J. H. (2012). Computational design of enhanced learning protocols. *Nature Neuroscience, 15*(2), 294–297.

Zrenner, E. (2002). Will retinal implants restore vision? [Research Support, Non-U.S. Gov't Review]. *Science, 295*(5557), 1022–1025.

INDEX

Note: Figures and tables are denoted with italicized page numbers; endnote information is denoted with an n and note number following the page number.